Introduction to Electronic Engineering

Introduction to Electronic Engineering

Josephine Wilson

Clanrye International
www.clanryeinternational.com

Clanrye International,
750 Third Avenue, 9th Floor,
New York, NY 10017, USA

ISBN: 978-1-64726-130-6

Cataloging-in-Publication Data

Introduction to electronic engineering / Josephine Wilson.
p. cm.
Includes bibliographical references and index.
ISBN 978-1-64726-130-6
1. Electronics. 2. Engineering. I. Wilson, Josephine.
TK7835 .I58 2022
621.381--dc23

For information on all Clanrye International publications
visit our website at www.clanryeinternational.com

CLANRYE
INTERNATIONAL

Contents

Preface

Electronics are concerned with the flow, emission and control of electrons in matter and vacuum. It deals with its engineering aspects as well as their applications. Electronic devices primarily contain an electronic circuit which consists of active and passive electrical componenets. Electonic engineering deals with the use of active and nonlinear electrical components in order to design VLSI devices, electronic circuits and other electronic devices. Some of the commonly used electrical components in this field include transistors, integrated circuits and diodes. Electronics engineering also facilitates the implementation of the principles and algorithms developed in areas such as signal processing, telecommunications, computer engineering, etc. This book studies, analyses and uphold the pillars of electronic engineering and its utmost significance in modern times. Also included in this book is a detailed explanation of the various concepts and applications of this domain. This textbook is an essential guide for both academicians and those who wish to pursue this discipline further.

A detailed account of the significant topics covered in this book is provided below:

Chapter 1- Electronic engineering is a subset of electrical engineering that uses nonlinear and active electrical components for designing devices, circuits, VLSI devices and their systems. It is primarily used in signal processing, telecommunication and information processing. This is an introductory chapter which will introduce briefly all the significant aspects of electronic engineering.

Chapter 2- An electronic component is a device that is used to affect electrons and their associated fields in an electronic system. Electronic Components can be categorized as passive, active or electromechanic. It includes components like inductors, resistors, diodes, amplifiers and capacitors. This chapter has been carefully written to provide an easy understanding of the varied facets of electronic components.

Chapter 3- A semiconductor device that is used to amplify signals and electronic powers is termed as a transistor. It is made up of semiconductor material with at least three terminals for connection to an external circuit. Field-effect transistor and bipolar junction transistor are the two main types of transistors. The diverse applications of transistors in the current scenario have been thoroughly discussed in this chapter.

Chapter 4- The electronic circuit is a circuit that is made up of individual electronic components that are connected by conductive wires through which electric current can flow. Analog circuits and digital circuits are the two primary types of electronic circuits used across all electronic devices. The chapter closely examines the key concepts of electronic circuits to provide an extensive understanding of the subject.

Chapter 5- Electronic systems with a continuous variable signal are known as analog electronics. It includes operational amplifiers, analog multipliers, instrumentation amplifiers, etc. Digital electronics are the electronic systems that majorly operate on digital signals. It is made up of large assemblies of logic gates. The topics elaborated in this chapter will help in gaining a better perspective about the analog and digital electronics.

It gives me an immense pleasure to thank our entire team for their efforts. Finally in the end, I would like to thank my family and colleagues who have been a great source of inspiration and support.

Josephine Wilson

1

Understanding Electronic Engineering

Electronic engineering is a subset of electrical engineering that uses nonlinear and active electrical components for designing devices, circuits, VLSI devices and their systems. It is primarily used in signal processing, telecommunication and information processing. This is an introductory chapter which will introduce briefly all the significant aspects of electronic engineering.

Electronics

Electronics is the branch of science that deals with the study of the flow of electrons and their effects in a vacuum, gases, and semiconductor materials. Electronics deals with electrical circuits that involve active and passive components such as vacuum tubes, transistors, diodes, integrated circuits, and sensors. Commonly, electronic devices contain circuitry consisting of active semiconductors with passive elements; such type of circuit is known as the electronic circuit.

Analog and Digital Electronics

Digital technology: Large digital clocks like this are quick and easy for runners to read

There are two very different ways of storing information—known as analog and digital. It sounds like quite an abstract idea, but it's really very simple. Suppose you take an old-fashioned photograph of someone with a film camera. The camera captures light streaming in through the shutter at the front as a pattern of light and dark areas on chemically treated plastic. The scene you're photographing is converted into a kind of instant, chemical painting—an "analogy" of what you're looking at. That's why we say this is an analog way of storing information. But if you take a photograph of exactly the same scene with a digital camera, the camera stores a very different record. Instead of saving a recognizable pattern of light and dark, it converts the light and dark areas into numbers and stores those instead. Storing a numerical, coded version of something is known as digital.

Electronic equipment generally works on information in either analog or digital format. In an old-fashioned transistor radio, broadcast signals enter the radio's circuitry via the antenna sticking out of the case. These are analog signals: they are radio waves, traveling through the air from a distant radio transmitter, that vibrate up and down in a pattern that corresponds exactly to the words and music they carry. So loud rock music means bigger signals than quiet classical music. The radio keeps the signals in analog form as it receives them, boosts them, and turns them back into sounds you can hear. But in a modern digital radio, things happen in a different way. First, the signals travel in digital format—as coded numbers. When they arrive at your radio, the numbers are converted back into sound signals. It's a very different way of processing information and it has both advantages and disadvantages. Generally, most modern forms of electronic equipment (including computers, cell phones, digital cameras, digital radios, hearing aids, and televisions) use digital electronics.

Electronic Components

If you've ever looked down on a city from a skyscraper window, you'll have marveled at all the tiny little buildings beneath you and the streets linking them together in all sorts of intricate ways. Every building has a function and the streets, which allow people to travel from one part of a city to another or visit different buildings in turn, make all the buildings work together. The collection of buildings, the way they're arranged, and the many connections between them is what makes a vibrant city so much more than the sum of its individual parts.

The circuits inside pieces of electronic equipment are a bit like cities too: they're packed with components (similar to buildings) that do different jobs and the components are linked together by cables or printed metal connections (similar to streets). Unlike in a city, where virtually every building is unique and even two supposedly identical homes or office blocks may be subtly different, electronic circuits are built up from a small number of standard components. But, just like LEGO, you can put these components together in an infinite number of different places so they do an infinite number of different jobs.

These are some of the most important components you'll encounter:

Resistors

These are the simplest components in any circuit. Their job is to restrict the flow of electrons and reduce the current or voltage flowing by converting electrical energy into heat. Resistors come in many different shapes and sizes. Variable resistors (also known as potentiometers) have a dial control on them so they change the amount of resistance when you turn them. Volume controls in audio equipment use variable resistors like these.

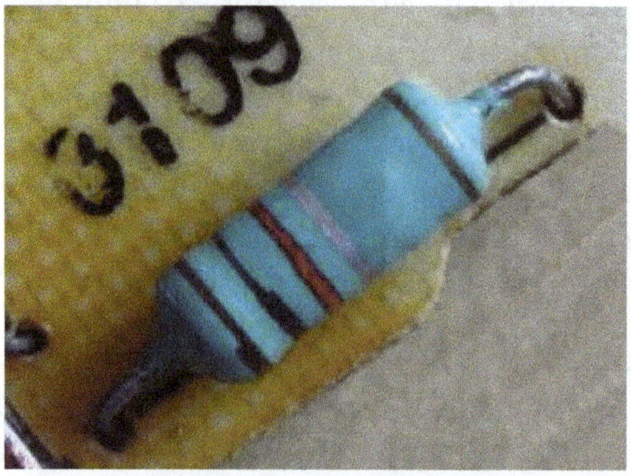

A typical resistor on the circuit board from a radio

Diodes

The electronic equivalents of one-way streets, diodes allow an electric current to flow through them in only one direction. They are also known as rectifiers. Diodes can be used to change alternating currents (ones flowing back and forth round a circuit, constantly swapping direction) into direct currents (ones that always flow in the same direction).

Diodes look similar to resistors but work in a different way and do a completely different job. Unlike a resistor, which can be inserted into a circuit either way around, a diode has to be wired in the right direction (corresponding to the arrow on this circuit board)

Capacitors

These relatively simple components consist of two pieces of conducting material (such as metal) separated by a non-conducting (insulating) material called a dielectric. They are often used as timing devices, but they can transform electrical currents in other ways too. In a radio, one of the most important jobs, tuning into the station you want to listen to, is done by a capacitor.

A small capacitor in a transistor radio circuit

Transistors

Easily the most important components in computers, transistors can switch tiny electric currents on and off or amplify them (transform small electric currents into much larger ones). Transistors that work as switches act as the memories in computers, while transistors working as amplifiers boost the volume of sounds in hearing aids. When transistors are connected together, they make devices called logic gates that can carry out very basic forms of decision making. (Thyristors are a little bit like transistors, but work in a different way.)

A typical field-effect transistor (FET) on an electronic circuit board

Opto-electronic (Optical Electronic) Components

There are various components that can turn light into electricity or vice-versa.Photocells (also known as photoelectric cells) generate tiny electric currents when light falls on them and they're used as "magic eye" beams in various types of sensing equipment, including some kinds of smoke detector. Light-emitting diodes (LEDs) work in the opposite way, converting small electric currents into light. LEDs are typically used on the instrument panels of stereo equipment. Liquid crystal displays (LCDs), such as those used in flatscreen LCD televisions and laptop computers, are more sophisticated examples of opto-electronics.

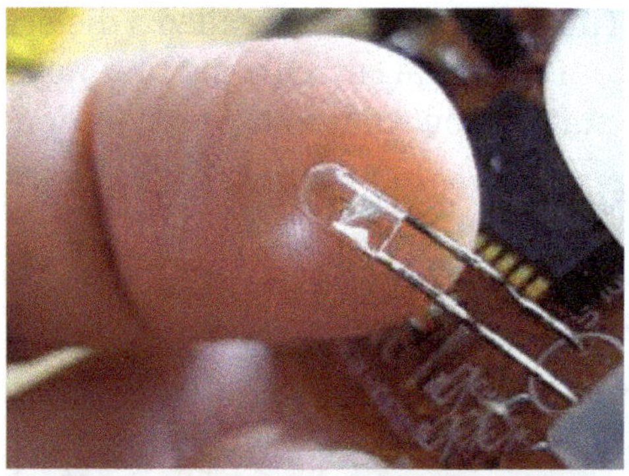

An LED mounted in an electronic circuit. This is one of the LEDs that makes red light inside an optical computer mouse

Electronic components have something very important in common. Whatever job they do, they work by controlling the flow of electrons through their structure in a very precise way. Most of these components are made of solid pieces of partly conducting, partly insulating materials called semiconductors. Because electronics involves understanding the precise mechanisms of how solids let electrons pass through them, it's sometimes known as solid-state physics. That's why you'll often see pieces of electronic equipment described as "solid-state."

Electronic Circuits

The key to an electronic device is not just the components it contains, but the way they are arranged in circuits. The simplest possible circuit is a continuous loop connecting two components, like two beads fastened on the same necklace. Analog electronic appliances tend to have far simpler circuits than digital ones. A basic transistor radio might have a few dozen different components and a circuit board probably no bigger than the cover of a paperback book. But in something like a computer, which uses digital technology, circuits are much more dense and complex and include hundreds, thousands, or even millions of separate pathways. Generally speaking, the more complex the circuit, the more intricate the operations it can perform.

If you've experimented with simple electronics, you'll know that the easiest way to build a circuit is simply to connect components together with short lengths of copper cable. But the more components you have to connect, the harder this becomes. That's why electronics designers usually opt for a more systematic way of arranging components on what's called a circuit board. A basic circuit board is simply a rectangle of plastic with copper connecting tracks on one side and lots of holes drilled through it. You can easily connect components together by poking them through the holes and using the copper to link them together, removing bits of copper as necessary, and adding extra wires to make additional connections. This type of circuit board is often called "breadboard".

Electronic equipment that you buy in stores takes this idea a step further using circuit boards that are made automatically in factories. The exact layout of the circuit is printed chemically onto a plastic board, with all the copper tracks created automatically during the manufacturing process. Components are then simply pushed through pre-drilled holes and fastened into place with a kind of electrically conducting adhesive known as solder. A circuit manufactured in this way is known as a printed circuit board (PCB).

Soldering components into an electronic circuit

The smoke you can see comes from the solder melting and turning to a vapor. The blue plastic rectangle being soldered onto here is a typical printed circuit board—and you see various components sticking up from it, including a bunch of resistors at the front and a large integrated circuit at the top.

Although PCBs are a great advance on hand-wired circuit boards, they're still quite difficult to use when you need to connect hundreds, thousands, or even millions of components together. The reason early computers were so big, power hungry, slow, expensive, and unreliable is because their components were wired together manually in this old-fashioned way. In the late 1950s, however, engineers Jack Kilby and Robert Noyce independently developed a way of creating electronic components in miniature form on the surface of pieces of silicon. Using these integrated circuits, it rapidly became possible to squeeze hundreds, thousands, millions, and then hundreds of millions of miniaturized components onto chips of silicon about the size of a finger nail. That's how computers became smaller, cheaper, and much more reliable from the 1960s onward.

Electric Current

An electric current is a flow of electric charge in a circuit. More specifically, the electric current is the rate of charge flow past a given point in an electric circuit. The charge can be negatively charged electrons or positive charge carriers including protons, positive ions or holes.

The magnitude of the electric current is measured in coulombs per second, the common unit for this being the Ampere or amp which is designated by the letter 'A'.

The Ampere or amp is widely used within electrical and electronic technology along with the multipliers like milliamp (0.001A), microamp (0.000001A), and so forth.

Current flow in a circuit is normally designated by the letter 'I', and this letter is used in equations like Ohms law where $V=I\cdot R$.

The basic concept of current is that it is the movement of electrons within a substance. Electrons are minute particles that exist as part of the molecular structure of materials. Sometimes these electrons are held tightly within the molecules and other times they are held loosely and they are able to move around the structure relatively freely.

Electrons are charged particles - they carry a negative charge. If they move then an amount of charge moves and this is called current.

It is also worth noting that the number of electrons that able to move governs the ability of a particular substance to conduct electricity. Some materials allow current to move better than others.

The motion of the free electrons is normally very haphazard - it is random - as many electrons move in one direction as in another and as a result there is no overall movement of charge.

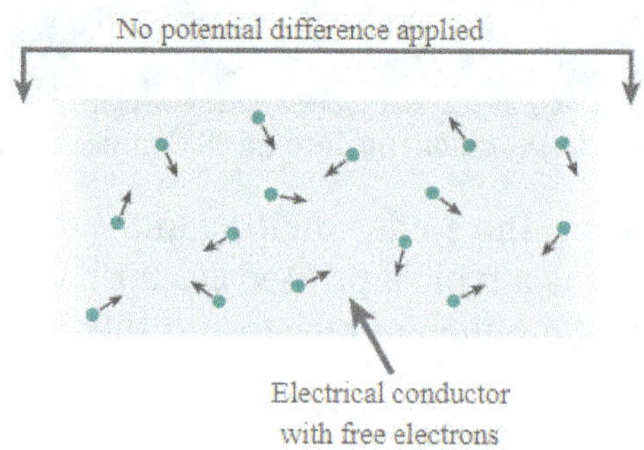

Random electron movement in a conductor with free electrons.

If a force acts on the electrons to move them in a particular direction, then they will all drift in the same direction, although still in a somewhat haphazard fashion, but there is an overall movement in one direction.

The force that acts on the electrons is called and electromotive force, or EMF, and its quantity is voltage measured in volts.

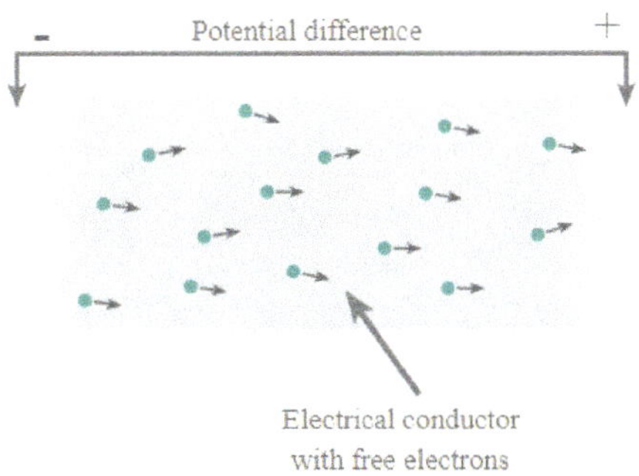

Electron flow under the action of applied electro-motive force.

To gain a little more understanding about what current is and how it acts in a conductor, it can be compared to water flow in a pipe. There are limitations to this comparison, but it serves as a very basic illustration of current and current flow.

The current can be considered to be like water flowing through a pipe. When pressure is placed on one end it forces the water to move in one direction and flow through the pipe. The amount of water flow is proportional to the pressure placed on the end. The pressure or force placed on the end can be likened to the electro-motive force.

When the pressure is applied to the pipe, or the water is allowed to flow as a result of a tap being opened, then the water flows virtually instantaneously. The same is true for the electrical current.

Conventional Current and Electron Flow

The particles that carry charge along conductors are free electrons. The electric field direction within a circuit is by definition the direction that positive test charges are pushed. Thus, these negatively charged electrons move in the direction opposite the electric field.

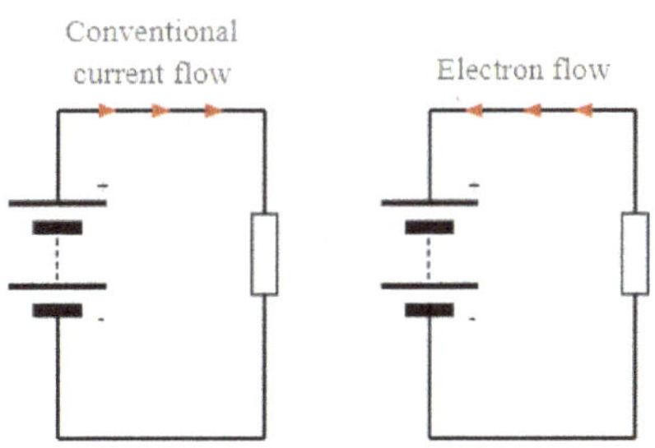

Electron and conventional current flow

This came about because the initial investigations in static and dynamic electric currents was based upon what we would now call positive charge carriers. This meant that then early convention for the direction of an electric current was established as the direction that positive charges would move.

- Conventional current flow: The conventional current flow is from positive to the negative terminal and indicates the direction that positive charges would flow.
- Electron flow: The electron flow is from negative to positive terminal. Electrons are negatively charged and are therefore attracted to the positive terminal as unlike charges attract.

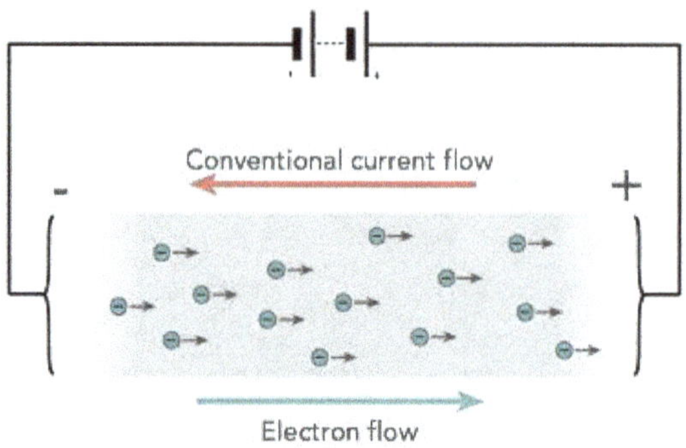

Speed of Electron or Charge Movement

The speed of the transmission of electrical current is very different to that of the speed of the actual electron movement. The electron itself bounces around in the conductor, and possibly only makes progress along the conductor at the rate of a few millimetres a second. This means that in the case of alternating current, where the current changes direction 50 or 60 times per second, most of the electrons never make it out of the wire.

Effects of Current

When an electric current flows through a conductor there are a number of signs which tell that a current is flowing:

- Heat is dissipated: Possibly the most obvious is that heat is generated. If the current is small then the amount of heat generated is likely to be very small and may not be noticed. However if the current is larger then it is possible that a noticeable amount of heat is generated. An electric fire is a prime example showing how a current causes heat to be generated. The actual amount of heat is governed not only be the current, but also be the voltage and the resistance of the conductor.
- Magnetic effect: Another effect which can be noticed is that a magnetic field is built up around the conductor. If a current is flowing in conductor then it is possible to detect this. By placing a compass close to a wire carrying a reasonably large direct current, the compass needle can be seen to be deflect. too fast for the needle to respond and the two wires (live and neutral) close together in the same cable will cancel out the field.

The magnetic field generated by a current is put to good use in a number of areas. By winding a wire into a coil, the effect can be increased, and an electro-magnet can be made. Relays and a host of other items use the effect. Loudspeakers also use a varying current in a coil to cause vibrations to occur in a diaphragm which enable the electronic currents to be converted into sounds.

Alternating Current

The current, which reverses its direction periodically in second, is called “Alternating Current”. The polarity of the voltage, is reversed periodically in an Alternating Current source. So the current reverses its direction periodically. Hence the name Alternating Current (AC) indicating current is alternating (periodically).

Number of times this current changes its direction in one second can be defined as the frequency of the alternating current. 50Hz frequency means it changes it direction for 50 times in a second.

In this current, the charged particles move from zero to maximum value in a direction and fall to zero and goes for another cycle in the opposite direction with the same values (in negative). The values processed by AC current in both the directions are equal at all instances.

It has both magnitude, direction and it varies in accordance with time. AC will flow in both positive and negative directions and hence it is Bi-directional. Generally AC waveform is represented by Sinusoidal waveform and its mathematical formula is:

$$A(t) = A \sin(2\pi f t)$$

Where,

A is magnitude of signal,

t is the time period,

f is the frequency of signal.

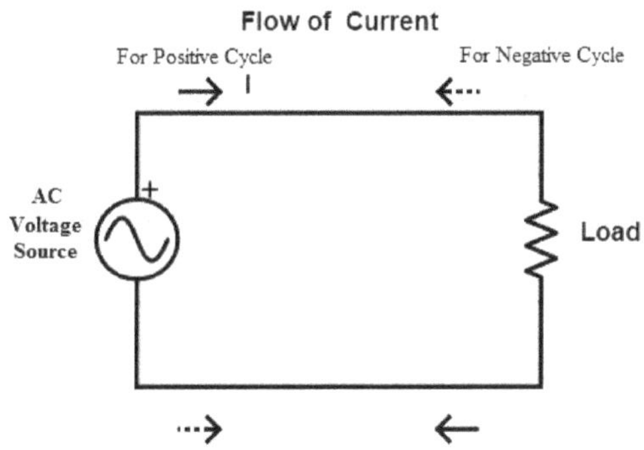

Generation of Alternating Current

Alternating current generating devices are called Alternators. The alternating current can be generated by many methods and by using many circuits. The simplest method of generating AC current is 'Using Basic Single Coil AC Generator'.

Working Principle

An AC generator works depending upon the principle of Faraday's electromagnetic induction to convert mechanical energy (such as rotation) into electrical energy.

Construction

A simple generator consists of two permanent magnets. They will produce a fixed magnetic field between north & south poles. In this magnetic field, there will be a single rectangular loop of wire which is an important part of the alternator.

It will rotate around a fixed axis in the magnetic field generated by the pole magnets and it cuts the magnetic flux. The central axis is perpendicular to the magnetic field generated by the two poles.

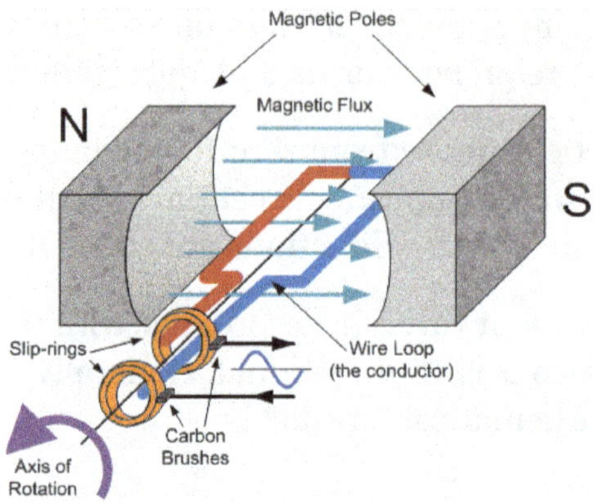

Process of Alternating Current Generation

- In the beginning, the wire is parallel to the field and does not cut magnetic flux i.e. the lines of magnetic force. So there is no voltage induced in the loop.
- As the wire rotates in a counter-clockwise direction, the wire sides will cut the magnetic lines of force in opposite directions.
- The induced EMF in the loop at any instant of time is proportional to the angle of rotation of the wire loop.
- When the wire loop has rotated for 180° across the magnetic lines of force in the opposite direction, then electrons in the wire flow in the opposite direction.
- This means, for every complete revolution (360°) of wire, one full sinusoidal waveform is produced.
- The rotating wire in magnetic field always connected with the carbon brushes and slip rings in all times.
- These are used to transfer the electrical current induced in the coil.
- The amount of EMF induced in a coil is determined by its speed, strength, and length.
- We know that 'frequency' is Number of cycles per second.
- Number of cycles produced will depend upon the speed of the rotating wire. That means the more the speed of the coil, the more will be the frequency. That means f is directly proportional to n, where 'n' number of revolutions of wire.
- The periodic reversal of polarity results in the generation of a voltage that has alternating polarity and hence the alternating current.

AC Wave Form

The alternating current can be represented by a waveform, with its amplitude and time period. There are many wave forms with which we can represent the AC current like square wave or triangular wave, but the easiest is to represent it with a sine wave.

For every complete cycle of the sine wave, we see a positive half and a negative half cycle. These positive and negative half cycles will represent that the AC current will change its direction periodically.

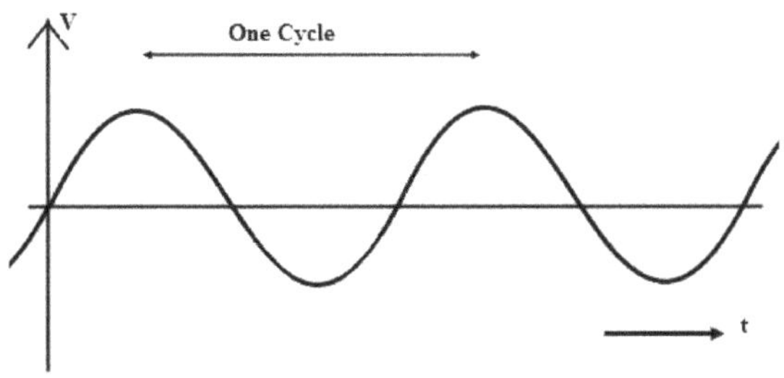

Mains (Line) Supply

The general purpose electric power supplied to households is called Mains electricity. The current in this supply is alternating current. The supply which is connected from the transformer power lines (Step up or step down) to the domestic power lines is called “Main supply or Line supply”. The main supply we use for home purpose in India is 220-240 volts with 50Hz frequency whereas in the US (United States) it is 120 volts with 60Hz frequency. The manufacturers of electrical equipment will design their products according to the Mains supply so that it will work for the required range of voltage and frequency.

Advantages of AC over DC

- If we consider the cost of generating AC and DC currents, the cost of AC current generation is cheaper than that of DC current generation.
- Machines using Alternating current are simple in design and their maintenance is also easy compared to machines using DC current.
- We can use the AC current for long distance transmission, but not DC because the voltage range of alternating current is high.
- AC current can be easily transformed from low to high voltage or high to low voltage by using a transformer. Hence the range of voltages in AC current is very wide.
- Without any loss we can convert the AC current to DC by using rectifiers and we can limit the magnitude of AC current with the help of an inductor or conductor at a very considerable loss.
- Long distance transmission losses in AC current are very less compared to DC transmission, when the supplied voltage is high.
- AC motors will produce a higher output (higher torque) than DC motors.

Disadvantages of AC over DC

- The peak value of AC current is high, so it is very dangerous to use without insulation.

- AC transmission cost is very high compared to DC transmission.
- It a gives repelling shock of high voltage to a person who touches it and can cause electrocution.
- There is a chance of interference with communication signals.
- Applications of AC current.
- All the electrical appliances we use in our day to day life work with AC current supply only. Ex: T.V., refrigerator, bulb, fan, Air conditioner etc.
- Alternating current is used in equipments where high power signals are transmitted. We use the Alternating Current in high power distribution systems like Antenna wave propagation, radio waves and ultrasonic wave propagation.
- We also use AC current in transformers for long distance transmission, to supply for many other purposes.
- Mainly we use the type of current in industries where the power requirement is high.

Direct Current

DC stands for Direct Current, which is electrical current that flows in one direction.

In dc circuits, the current is in one direction unlike the alternating current (AC) where the current reverses direction 50 or 60 times a second depending on the frequency of the supply. As the direct current flows, the electrons, which constitute the electric charge, flow from the point of low potential to the point of high potential. They move from the negative terminal to the positive terminal and the resulting current is in the opposite direction (from positive to negative).

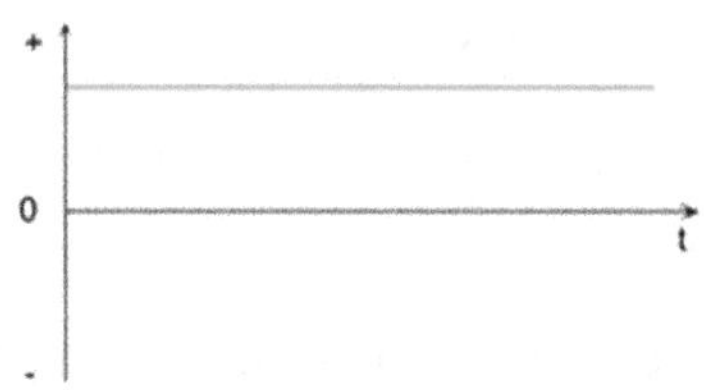

Plot of direct current/voltage against time

The DC is usually used in low-voltage applications such as battery operated equipment. When a DC is required, and the use of battery is uneconomical or requires constant charging, a power supply is used to convert the ac to dc. This will charge the battery, and the equipment used off the mains until the battery is depleted (i.e. a mobile phone). The other option is to power the circuit directly from the rectified DC and have the battery work as a backup power when the AC mains supply is unavailable (i.e. in a laptop).

Power supplies range from simple regulators with one output, to smps power supplies that are more regulated and capable of supplying multiple outputs such as those required to power a computer. The voltages and current's capacity of a power supply depend on the design and components used. There are a variety of DC voltage levels, typical values are 1.2, 1.5, 3, 3.3, 3.6, 5, 6, 10, 12, 15, 18, 18.5, 19, 20, 24, and 48 Volts.

An ideal DC supply should provide a constant voltage and adequate current throughout the operation of the equipment. However, sources such as batteries have a limited capacity and can only power the equipment efficiently for a given period determined by the rating of the battery and the load.

To maintain the power at a constant level, most equipment uses rechargeable batteries so that the charge can be replenished regularly. The charger consists of a rectifier circuit which converts the readily available AC into a suitable DC voltage.

Other than changing the mains AC to DC, most power supplies change the level of the voltage. Most of them will always lower the voltage since most of the electronics functions from much less voltage levels, but higher currents. Most laptops use between 18 Volts DC to 20 V DC and current of at least 3A. The laptop adapter must therefore provide a means of lowering the mains voltage from either the 220V or 120 Volts down to about 20 Volts DC.

Sources of Direct Current

- DC generators
- Batteries
- DC power converters that rectify the AC
- Solar panels
- Thermocouples.

Advantages of DC

- Most equipment are more efficient when directly powered from DC
- More efficient especially when the power leads are less than 100 feet
- Less risks of shock hazard for voltages below 48 Volts.

Disadvantages of DC

- High cost when dealing with high current systems; i.e. larger cables, fuses, switches and other components required for high current, low voltage applications
- Difficult to obtain DC appliances and equipment
- Higher chances of fire hazards when maximum circuit ratings and wire sizes are not strictly adhered to.

Applications

DC is used in almost all electronics equipment, electric vehicles, automation, control of electrical equipment, and more. Most office and domestic equipment such as TVs, audio systems, amplifiers, flash lights, computers, tablets and smartphones use DC power to function. However, since the universally available power is AC, the equipment uses an external or internal power supply to convert the utility mains AC into the desired DC for the equipment.

Voltage

Voltage is the electromotive force or the electric potential energy difference between two points (often within the context of an electrical circuit) per unit of charge, expressed in volts (V). Voltage, along with current and resistance, is used to describe the behavior of electrons.

Voltage is one of the fundamental parameters associated with any electrical or electronic circuit. Voltage is seen widely in specifications of a host of electrical items from batteries to radios and light bulbs to shavers, and on top of this it is a key parameter that is measured within circuits as well.

The operating voltage of an item of equipment is very important - it is necessary to connect electrical and electronic items to supplies of the correct voltage. Connect a 240 volt light bulb to a 12 volt battery and it will not light up, but connect a small 5V USB device to a 240 volt supply and far too much current will flow and it will burn up and be irreparably damaged.

On top of this, the voltage levels within a circuit give a key to its operation - if the incorrect voltage is present, then it may give an indication of the reason for the malfunction.

For these and many reasons, electrical voltage is a key parameter and knowing what the voltage is can be a key requirement in any circumstance.

Voltage can be considered as the pressure that forces the charged electrons to flow in an electrical circuit. This flow of electrons is the electrical current that flows.

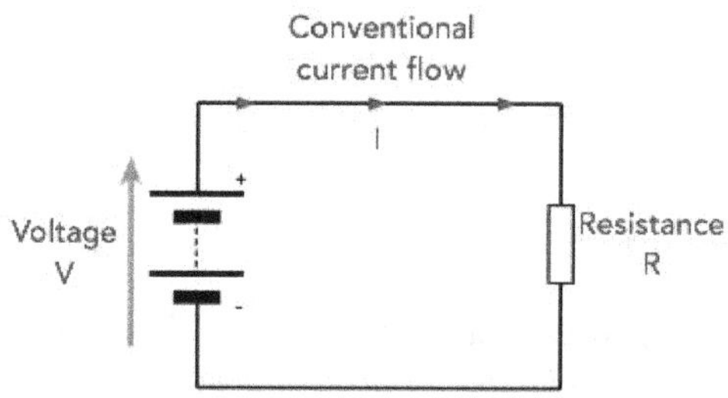

Voltage shown within a simple circuit.

If a positive potential is placed on one end of a conductor, then this will attract that negative charges to it because unlike charges attract. The higher the potential attracting the charges, the greater the attraction and the greater the current flow.

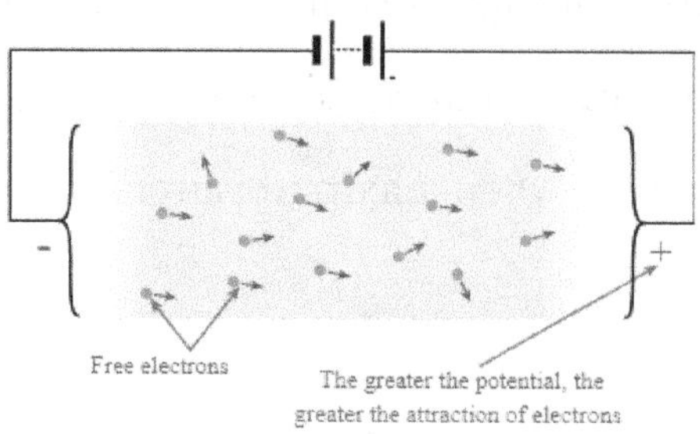

The higher the voltage potential difference, the greater the attraction of electrons and greater the current flow.

In essence, the voltage is the electrical pressure and it is measured in volts which can be represented by the letter V.

Normally the letter V is used for volts in an equation like Ohm's law, but occasionally the letter E may be used - this stands for EMF or electro-motive force.

Potential Difference

The electrical potential or voltage is a measure of the electrical pressure available to force the current around a circuit. In the comparison of a water system mentioned when describing current, the potential can be likened to the water pressure at a given point. The greater the pressure difference across a section of the system, the greater the amount of water which will flow. Similarly the greater the potential difference or voltage across a section of an electrical circuit, the greater the current which will flow.

Volt

The basic unit of voltage is the volt, named after the Italian scientist, Alessandro Volta, who made some early batteries and performed many other experiments with electricity.

To give an idea of the voltages which are likely to be encountered, a CB radio will usually operate from a supply of around 12 volts (12 V). The cells used in domestic batteries have a voltage of around 1.5 volts. Rechargeable Nickel Cadmium cells have a slightly smaller voltage of 1.2 volts, but can normally be used interchangeably with the non-rechargeable types.

In other areas voltages much smaller and much greater than this can be encountered. The signal input to an audio amplifier will be smaller than this, and the voltages will often be measured in millivolts (mV) or thousandths of a volt. The signals at the input to a radio are even smaller than this and will often be measured in microvolts (μV) or millionths of a volt.

The cathode ray tubes in a television or computer monitors require voltages of several kilovolts (kV) or thousands of volts, and even larger voltages of millions of volts or megavolts (MV).

EMF and PD

The voltage for a battery or single cell is stated as a voltage. However it is found that when the battery is in use its voltage will fall, especially as it becomes older and it has been used. The reason for this is that there is some resistance inside the cell. As current flows a voltage drop forms across this and the voltage seen at the output is less than expected. Even so the voltage seen at the terminals if the battery was not supplying current would still be the same. This no load voltage is known as the electro-motive force (EMF), and is the internal voltage which is generated by the cell, or other source of power.

Ohm's Law

Ohm's law states that the voltage or potential difference between two points is directly proportional to the current or electricity passing through the resistance, and inversely proportional to the resistance of the circuit. The formula for Ohm's law is $V=I/R$. This relationship between current, voltage, and relationship was discovered by German scientist Georg Simon Ohm.

Most basic components of electricity are voltage, current, and resistance. Ohm's law shows a simple relation between these three quantities. Ohm's law states that the current through a conductor between two points is directly proportional to the voltage across the two points.

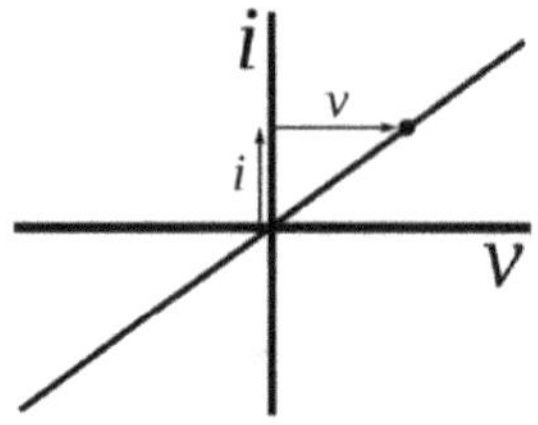

Ohm's Law Formula

Voltage= Current× Resistance

V= I×R

V= voltage, I= current and R= resistance

The SI unit of resistance is ohms and is denoted by Ω.

This law is one of the most basic laws of electricity. It helps to calculate the power, efficiency, current, voltage, and resistance of an element of an electrical circuit.

Applications of Ohm's Law

Ohm's law helps us in determining either voltage, current or impedance or resistance of a linear electric circuit when the other two quantities are known to us. It also makes power calculation simpler.

Relationship between Current-Voltage

In order to establish the current-voltage relationship, the ratio V / I remains constant for a given resistance, therefore a graph between the potential difference(V) and the current (I) must be a straight line.

Finding the Values of Resistance

It is the constant ratio that gives the unknown values of resistance,

$$\frac{V}{I} = R$$

For a wire of uniform cross-section, the resistance depends on the length l and the area of cross-section A. It also depends on the temperature of the conductor. At a given temperature the resistance,

$$R = \rho \frac{1}{A}$$

where ρ is the specific resistance or resistivity and is characteristic of the material of wire. The specific resistance or resistivity of the material of the wire is,

$$\rho = \frac{RA}{1}$$

If 'r' is the radius of the wire, then the cross-sectional area, $A = \pi r^2$. Then the specific resistance or resistivity of the material of the wire is,

$$\rho = \frac{\pi r^2 R}{1}$$

Limitations of Ohms Law

1. Ohm's law is not applicable to unilateral networks. Unilateral networks allow the current to flow in one direction. Such types of network consist elements like a diode, transistor, etc.
2. Ohm's law is also not applicable to non – linear elements. Non-linear elements are those which do not have current exactly proportional to the applied voltage that means the resistance value of those elements changes for different values of voltage and current. Examples of non – linear elements are the thyristor.

Advantages of Electronics

Electronic devices are playing a major role in everyday life. The various electronic devices we use in everyday life include:

- Computers

  Today, computers are using everywhere. At home, computers are used for playing games, watching movies, doing research, paying bills and reservation of tickets for railways and airlines. At school, students use computers to complete their assignments.

- Mobile phones

  Mobile phones are used for variety of purposes such as for sending text messages, making voice calls, surfing internet, playing games, and listening songs.

- ATM

  ATM is an electronic telecommunication device particularly used for withdrawing money at any time from anywhere. ATM stands for automated teller machine. The customer can withdraw money up to a certain limit during anytime of the day or night.

- Pen drive

  Pen drive is particularly used for storing large amount of data and also used for transferring data from one device to another. For example, the data stored in the computer can be transferred to the pen drive. The data stored in this pen drive can be retrieved at anytime.

- Television

  Television is an electronic device primarily used for entertainment and knowledge. It is used for watching movies for entertainment, news for knowledge, cartoons for children's.

- Digital camera

  Digital camera is a camera used for taking pictures and videos. This images and videos are stored for later reproduction.

Electronic Engineering

Electronic engineering, or electronics engineering is a form of engineering associated with electronic circuits, devices and the equipment and systems that use them.

Electronic engineering utilises a variety of different types of electronic components from the more traditional analogue components through to digital electronic components, microprocessors and microcontrollers as well as programmable logic devices. This means that electronic engineering can incorporate a large variety of different areas.

The field of electronic engineering includes a variety more specific electronic engineering fields including: analogue electronics, digital electronics, consumer electronics, embedded systems and power electronics.

Electronic Engineering Fields

When asking the question: What is electronic Engineering, it is necessary to understand that there are many different fields and disciplines involved within the overall topic.

Normally electronic engineers specialise in a particular field or area. This means that it is possible to select a particular field of interest.

Some of the fields within electronic engineering include:

- Analogue electronic engineering: Analogue electronics is still a major sector within the overall electronic engineering scene. With many analogue elements still being needed, analogue circuits are still widely used. Whilst analogue engineering is not as large as it used to be many years ago before digital electronics took off in a big way, the growth in the overall electronics market has compensated for this. Often analogue electronic engineering provides a stimulating environment in which to work.
- Radio frequency engineering: radio frequency electronic engineering has grown in its size in recent years. With many more systems using wireless links, everything from mobile phones to Wi-Fi, IoT, short range links and very much more, wireless technology is needed. Accordingly radio frequency design is in great demand.
- Digital development engineering; : Many functions are now undertaken using digital techniques. Accordingly many digital circuits are needed and this means that some digital/logic electronic engineering is needed. That said the demand for digital engineers alone may not be as large as anticipated.
- Programmable logic engineering: With the complexity of many logic / digitally based circuits, an approach that is being used increasingly is one where programmable logic chips are used. FPGAs, and other programmable logic chips are widely used, enabling large amounts of logic to be incorporated into programmable chips. Using high levels design languages like VHDL, etc, the design is brought to within manageable limits. Also if the design needs optimising or changing, this can be achieved by changing the logic programme. This sector of the electronic engineering arena is growing, and the tools becoming more sophisticated and very interesting to use.

- Software engineering: There is an increasing amount of software contained within electronic products these days. As a result software engineering is becoming increasingly important. In many projects, at least two thirds of the development budget is allocated to developing the software – an indication of how this sector has grown.
- Systems engineering: Systems engineering is a particularly important element of the design of any item. In terms of this sector of electronic engineering, a system is any completed object. It may be a radio receiver, it may be a mobile phone, or it could be an item consisting of several individual items. The terms systems engineering refers to the fact that this form of engineering looks at the complete object or system, comprising of smaller items, everything from boards to complete units. It looks at the operation of the overall system, ensuing that the initial requirements are correct, and that the item is finally tested to ensure that it operates to its specification, and also to the initial requirements that were placed upon it.

Applications of Electronics

Aerospace Industry

Today with advancement in electronics engineering application in the field of aerospace has open many impossible ways to humans. Most of the parts used in the Space shuttle, Satellite power supplies, aircraft power management, and other communication instruments are electronics devices. Even in commercial airlines, there are numerous electronic devices which are used to measure different physical parameters like humidity, temperature, pressure, elevation, etc.

Medical

With the advancement of the electronics field, and particularly that of computer technology has made possible many of the application which made possible for nurses and doctors to examine his patient. Many machines like NMR, ECE, x-ray, etc use electronics. There is noticeable advancement of electronics in the field of medical sciences. There are a few mindblowing innovations which prove how important is electronics – Robotic Arm, Robotic Check-Ups, Bionic Eye, Needle-Free Diabetes Care, etc.

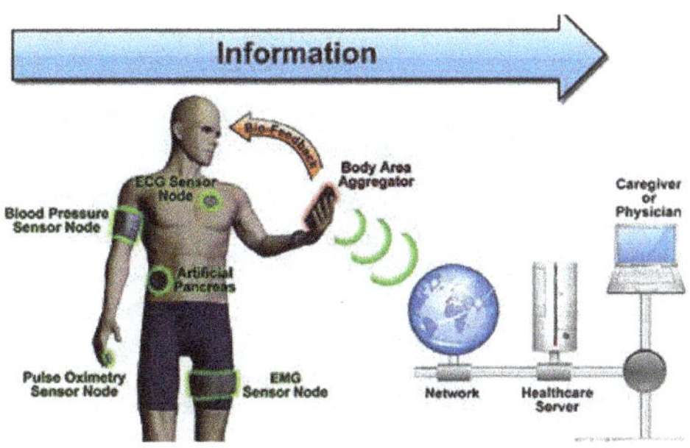

Automobile

Electronics are used in road vehicles, such as Lighting system of the car, telematics, in-car sound systems, etc. Initially, electronics were used to in cars to control engines. Nowadays almost every things are controlled using electronics whether it's a chair in car or speed limitation in car, or Gas management in the car. Now electronic with its advancement, it is used in following parts of cars:

1. Transmission
2. Engine
3. Chassis
4. Passenger comfort
5. Active safety Driver assistance
6. Entertainment systems

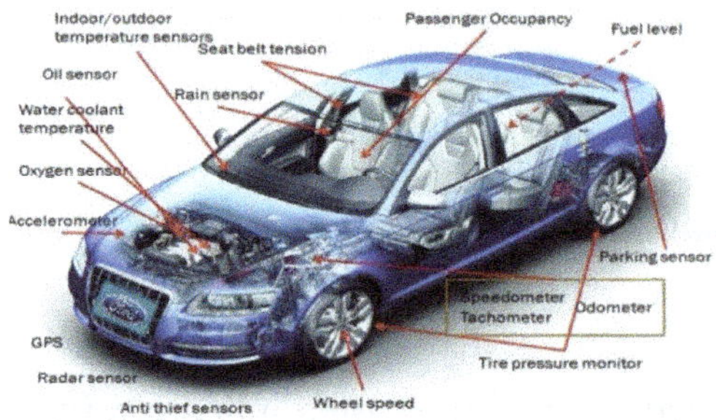

Utility Systems

Electronics has also applications in utility systems such as High voltage DC Transmission, VAR compensation, Static circuit breakers, Excitation systems, fans and boiler feed pumps, solar, wind power management system, Smart grid system.

Commercial

Electronic devices and gadgets are widely used for the commercial purpose such as Telephone,

Airconditioning, heating, Digital Advertising, power supplies, Personal computer, MP3s, office equipment, light dimmer, uninterruptible power supplies (UPS), etc.

Agriculture

Electronics has proved to be the biggest problem solver in the agriculture field. Today number of electronics devices and sensors are used to monitor a crop or enhance the production of Agriculture. Like e-Agri Sensors and other electronic gadgets which are used for measuring the salinity of soil, moisture level, and nutrition level in the soil.

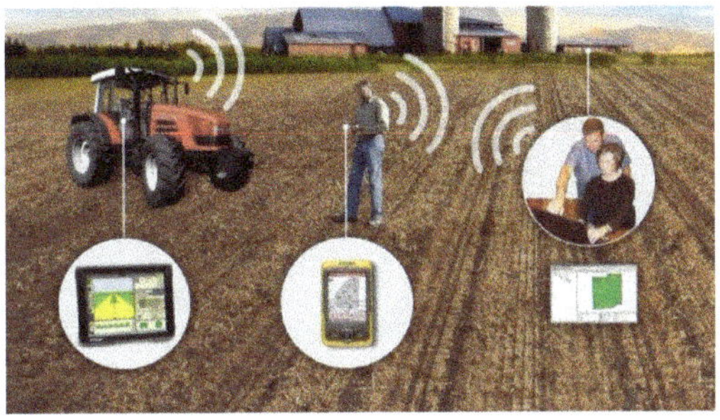

Communication

In order to communicate for far distance, Electronic devices and systems are used such as Am Radio, FM Radio, Television, processing, Memory storage, display, analysis, protection, etc.

Electronic sensors and devices are used to allow friendly forces to operate in the virtual environment. it is very useful we can play games such as Need for speed, PUBG, GTA, Prince of Persia, Counter-Strike.

Industrial

Electronics applications are widely utilized in Arc and Industrial furnaces such as blowers, fans, pumps and compressors, industrial lasers, transformer tap changers, rolling mills, textile mills, excavators, cement mills, welding. Nowadays most of the firms use robotics technology Such as in the manufacturing of cars.

Residential

Electronics devices which make our life easy and better such as air conditioner, cooking appliances, dryer, personal computer, mobile phones, etc.

Military

Electronics devices and machinery are also widely used in the military such as UAV and drones which are used in the military for aerial attack as well as for monitoring. Magnetic anomaly detector, People sniffer, Night vision device, infrared detector, RADAR, PNS Hameed, and night vision camera etc. are some gadgets used by the military.

References

- What-is-electronics: studentsheart.com, Retrieved 3 March, 2019
- Electronics: explainthatstuff.com, Retrieved 13 May, 2019
- What-is-electrical-current, current, basic_concepts: electronics-notes.com, Retrieved 6 February July, 2019
- Introduction-to-alternating-current: electronicshub.org, Retrieved 23 January, 2019
- What-is-dc-direct-current: sunpower-uk.com, Retrieved 25 April, 2019
- What-is-voltage-basics-tutorial, voltage, basic_concepts: electronics-notes.com, Retrieved 24 June, 2019
- Ohms-law-and-resistance, electricity, physics: toppr.com, Retrieved 19, 2019
- Electronic-devices-and-circuits: physics-and-radio-electronics.com, Retrieved 17 August, 2019
- What-is-electronic-engineering, electronics-engineering, becoming-an-engineer: electronics-notes.com, Retrieved 7 January, 2019
- What-is-electronics: studentsheart.com, Retrieved 5 April, 2019

2

Electronic Components and Devices

An electronic component is a device that is used to affect electrons and their associated fields in an electronic system. Electronic Components can be categorized as passive, active or electromechanic. It includes components like inductors, resistors, diodes, amplifiers and capacitors. This chapter has been carefully written to provide an easy understanding of the varied facets of electronic components.

Electronic components are the elements of circuit which helps in its functioning. They can be classified into two types i.e. Active Components and Passive Components. Active elements are those which possess gain. They can give energy to the circuit. On the contrary passive elements do not possess gain and they cannot give energy continuously to the circuit.

Passive components cannot amplify or energize the energy of the signal associated with them, they can only attenuate it, while active components can energize or amplify the signal. For example, transistors, amplifiers are active elements. While resistor, inductor, capacitor are passive elements as they cannot supply energy to the circuit.

Classsification of Electronic Components

Electronic component can be defined into various types based on various parameter such as measurement components, power supply components etc.

1. Elementary Components based on design architecture: Diodes, Transistors, resistors, capacitors, inductors, LED etc. These are the basic components of a circuit as without these components the circuit cannot not process or work properly.

2. Power Supply Components: DC power supplies, AC power supply, signal generators are the components which fall in this category. These are also the considered as active components as they supply energy to the circuit.

3. Measurement devices: Cathode ray oscilloscope and millimeters are usual measurement devices in electronic circuits.

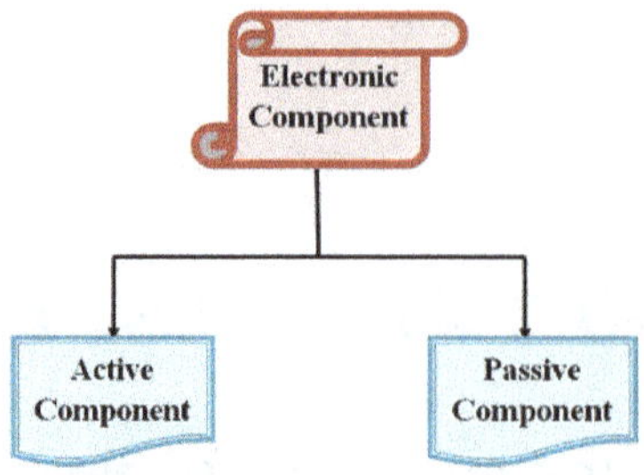

The classification of electronic Components.

Resistor

A resistor is an electrical component that limits or regulates the flow of electrical current in an electronic circuit. Resistors can also be used to provide a specific voltage for an active device such as a transistor.

All other factors being equal, in a direct-current (DC) circuit, the current through a resistor is inversely proportional to its resistance, and directly proportional to the voltage across it. This is the well-known Ohm's Law. In alternating-current (AC) circuits, this rule also applies as long as the resistor does not contain inductance or capacitance.

Resistors can be fabricated in a variety of ways. The most common type in electronic devices and systems is the carbon-composition resistor. Fine granulated carbon (graphite) is mixed with clay and hardened. The resistance depends on the proportion of carbon to clay; the higher this ratio, the lower the resistance.

Another type of resistor is made from winding Nichrome or similar wire on an insulating form. This component, called a wirewound resistor, is able to handle higher currents than a carbon-composition resistor of the same physical size. Because the wire is wound into a coil, the component acts as an inductors as well as exhibiting resistance. This does not affect performance in DC circuits, but can have an adverse effect in AC circuits because inductance renders the device sensitive to changes in frequency.

Resistors in Parallel

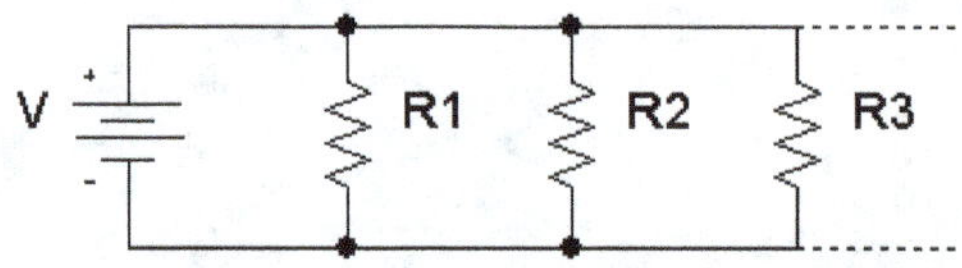

The total equivalent resistance of resistors in parallel R_{Total} is given by:

$$\frac{1}{R_{Total}} = \frac{1}{R_1} + \frac{1}{R_2} + \frac{1}{R_3} + \ldots$$

So when you add resistors in parallel, the total resistance is decreased.

Resistors in Series

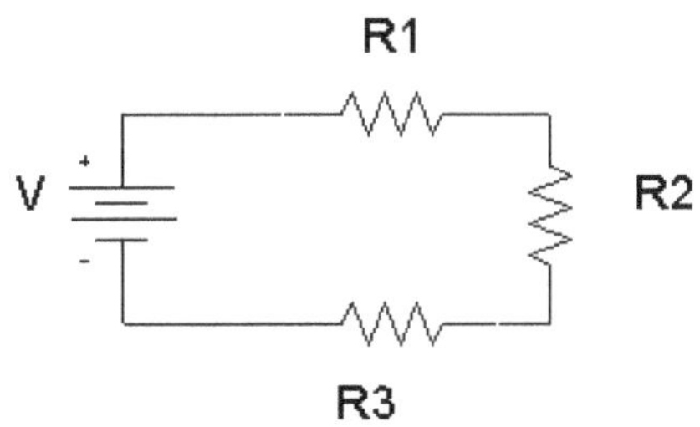

The total equivalent resistance of resistors in series R_{total} is the sum of the resistance values:

$$R_{total} = R_1 + R_2 + R_3 + ...$$

So when you add resistors in series, the total resistance is increased.

Dimensions and Material Affects

The resistance R in ohms (Ω) of a resistor is equal to the resistivity ρ in ohm-meters (Ω·m) times the resistor's length l in meters (m) divided by the resistor's cross sectional area A in square meters (m²):

$$R = \rho \times \frac{l}{A}$$

Resistor Symbols

	Resistor (IEEE)	Resistor reduces the current flow
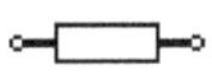	Resistor (IEC)	
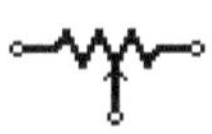	Potentiometer (IEEE)	Adjustable resistor - has 3 terminals
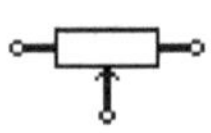	Potentiometer (IEC)	
	Variable Resistor / Rheostat (IEEE)	Adjustable resistor - has 2 terminals
	Variable Resistor / Rheostat (IEC)	

	Trimmer Resistor	Presest resistor
	Thermistor	Thermal resistor - change resistance when temperature changes
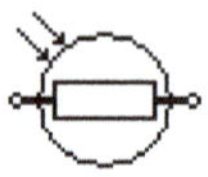	Photoresistor / Light dependent resistor (LDR)	Changes resistance according to light

Resistor Color Code

The resistance of the resistor and its tolerance are marked on the resistor with color code bands that denotes the resistance value.

There are 3 types of color codes:

- 4 bands: digit, digit, multiplier, tolerance.
- 5 bands: digit, digit, digit, multiplier, tolerance.
- 6 bands: digit, digit, digit, multiplier, tolerance, temperature coefficient.

Resistance calculation of 4 bands resistor:

$$R = (10 \times digit_1 + digit_2) \times multiplier$$

Resistance calculation of 5 or 6 bands resistor:

$$R = (100 \times digit_1 + 10 \times digit_2 + digit_3) \times multiplier$$

Types of Resistors

There are many different Types of Resistor available for the electronics constructor to choose from, from very small surface mount chip resistors up to large wirewound power resistors:

Composition Types of Resistor

Carbon Resistors are the most common type of Composition Resistors. Carbon resistors are a cheap general purpose resistor used in electrical and electronic circuits. Their resistive element is manufactured from a mixture of finely ground carbon dust or graphite (similar to pencil lead) and a non-conducting ceramic (clay) powder to bind it all together.

Carbon Resistor

The ratio of carbon dust to ceramic (conductor to insulator) determines the overall resistive value of the mixture and the higher the ratio of carbon, the lower the overall resistance. The mixture is moulded into a cylindrical shape with metal wires or leads are attached to each end to provide the electrical connection as shown, before being coated with an outer insulating material and colour coded markings to denote its resistive value.

Carbon Resistor

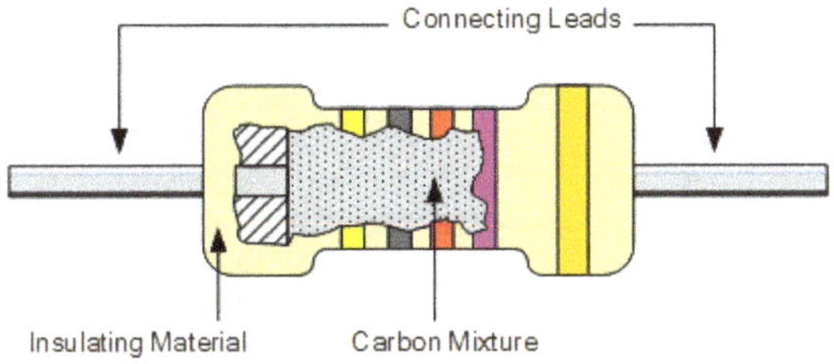

The Carbon Composite Resistor is a low to medium type power resistor which has a low inductance making them ideal for high frequency applications but they can also suffer from noise and stability when hot. Carbon composite resistors are generally prefixed with a "CR" notation (eg, CR10kΩ) and are available in E6 (± 20% tolerance (accuracy)), E12 (± 10% tolerance) and E24 (± 5% tolerance) packages with power ratings from 0.250 or 1/4 of a Watt up to 5 Watts.

Carbon composite resistor types are very cheap to make and are therefore commonly used in electrical circuits. Due to their manufacturing process carbon type resistors have very large tolerances so for more precision and high value resistances, film type resistors are used instead.

Film Type Resistors

The generic term "Film Resistor" consist of *Metal Film*, *Carbon Film* and *Metal Oxide Film*resistor types, which are generally made by depositing pure metals, such as nickel, or an oxide film, such as tin-oxide, onto an insulating ceramic rod or substrate.

Film Resistor

The resistive value of the resistor is controlled by increasing the desired thickness of the deposited film giving them the names of either "thick-film resistors" or "thin-film resistors".

Once deposited, a laser is used to cut a high precision spiral helix groove type pattern into this film. The cutting of the film has the effect of increasing the conductive or resistive path, a bit like taking a long length of straight wire and forming it into a coil.

This method of manufacture allows for much closer tolerance resistors (1% or less) as compared to the simpler carbon composition types. The tolerance of a resistor is the difference between the preferred value (i.e, 100 ohms) and its actual manufactured value i.e, 103.6 ohms, and is expressed as a percentage, for example 5%, 10% etc, and in our example the actual tolerance is 3.6%. Film type resistors also achieve a much higher maximum ohmic value compared to other types and values in excess of 10MΩ (10 Million Ohms) are available.

Film Resistor

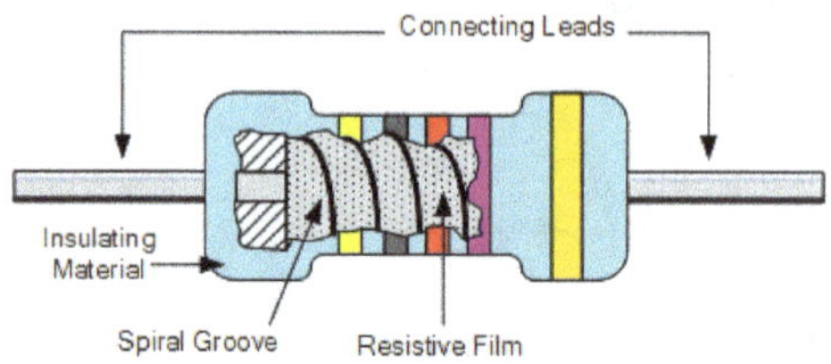

Metal Film Resistors have much better temperature stability than their carbon equivalents, lower noise and are generally better for high frequency or radio frequency applications. Metal Oxide Resistors have better high surge current capability with a much higher temperature rating than the equivalent metal film resistors.

Another type of film resistor commonly known as a Thick Film Resistor is manufactured by depositing a much thicker conductive paste of CERamic and METal, called Cermet, onto an alumina ceramic substrate. Cermet resistors have similar properties to metal film resistors and are generally used for making small surface mount chip type resistors, multi-resistor networks in one package for pcb's and high frequency resistors. They have good temperature stability, low noise, and good voltage ratings but low surge current properties.

Metal Film Resistors are prefixed with a "MFR" notation (eg, MFR100kΩ) and a CF for Carbon Film types. Metal film resistors are available in E24 (±5% & ±2% tolerances), E96(±1% tolerance) and E192 (±0.5%, ±0.25% & ±0.1% tolerances) packages with power ratings of 0.05 (1/20th) of a Watt up to 1/2 Watt. Generally speaking Film resistors and especially metal film resistors are precision low power components.

Wirewound Types of Resistor

Another type of resistor, called a Wirewound Resistor, is made by winding a thin metal alloy wire (Nichrome) or similar wire onto an insulating ceramic former in the form of a spiral helix similar to the film resistor.

Wirewound Resistor

These types of resistor are generally only available in very low ohmic high precision values (from 0.01Ω to 100kΩ) due to the gauge of the wire and number of turns possible on the former making them ideal for use in measuring circuits and Wheatstone bridge type applications.

They are also able to handle much higher electrical currents than other resistors of the same ohmic value with power ratings in excess of 300 Watts. These high power resistors are moulded or pressed into an aluminium heat sink body with fins attached to increase their overall surface area to promote heat loss and cooling.

These special types of resistor are called "Chassis Mounted Resistors" because they are designed to be physically mounted onto heatsinks or metal plates to further dissipate the generated heat. The mounting of the resistor onto a heatsink increases their current carrying capabilities even further.

Another type of wirewound resistor is the Power Wirewound Resistor. These are high temperature, high power non-inductive resistor types generally coated with a vitreous or glass epoxy enamel for use in resistance banks or DC motor/servo control and dynamic braking applications. They can even be used as low wattage space or cabinet heaters.

The non-inductive resistance wire is wound around a ceramic or porcelain tube covered with mica to prevent the alloy wires from moving when hot. Wirewound resistors are available in a variety of resistance and power ratings with one main use of power wirewound resistor is in the electrical heating elements of an electric fire which converts the electrical current flowing through it into heat with each element dissipating up to 1000 Watts, (1kW) of energy.

Because the wire of standard wire wound resistors is wound into a coil inside the resistors body, it acts like an inductor causing them to have inductance as well as resistance. This affects the way the resistor behaves in AC circuits by producing a phase shift at high frequencies especially in the larger size resistors. The length of the actual resistance path in the resistor and the leads contributes inductance in series with the "apparent" DC resistance resulting in an overall impedance path of Z Ohms.

Impedance (Z) is the combined effect of resistance (R) and inductance (X), measured in ohms and for a series AC circuit is given as, $Z^2 = R^2 + X^2$.

When used in AC circuits this inductance value changes with frequency (inductive reactance, $X_L = 2\pi f L$) and therefore, the overall value of the resistor changes. Inductive reactance increases with frequency but is zero at DC (zero frequency). Then, wirewound resistors must not be designed or used in AC or amplifier type circuits where the frequency across the resistor changes. Special non-inductive wirewound resistors are also available.

Wirewound Resistor

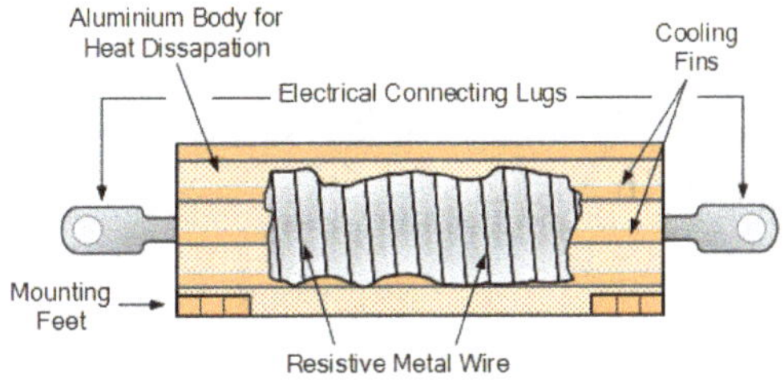

Wirewound resistor types are prefixed with a "WH" or "W" notation (eg WH10Ω) and are available in the WH aluminium clad package (±1%, ±2%, ±5% and ±10% tolerance) or the W vitreous enamelled package (±1%, ±2% and ±5% tolerance) with power ratings from 1W to 300W or more.

Capacitor

Capacitor is an electronic component that stores electric charge. The capacitor is made of 2 close conductors (usually plates) that are separated by a dielectric material. The plates accumulate electric charge when connected to power source. One plate accumulates positive charge and the other plate accumulates negative charge.

The capacitance is the amount of electric charge that is stored in the capacitor at voltage of 1 Volt.

The capacitance is measured in units of Farad (F).

The capacitor disconnects current in direct current (DC) circuits and short circuit in alternating current (AC) circuits.

Capacitor is a two-terminal passive component having property of capacitance. This property electrifies (charging with electricity) capacitor with input voltage. As capacitors condense (store) electricity, hence it is also known as condenser.

With this definition we can define capacitance as,

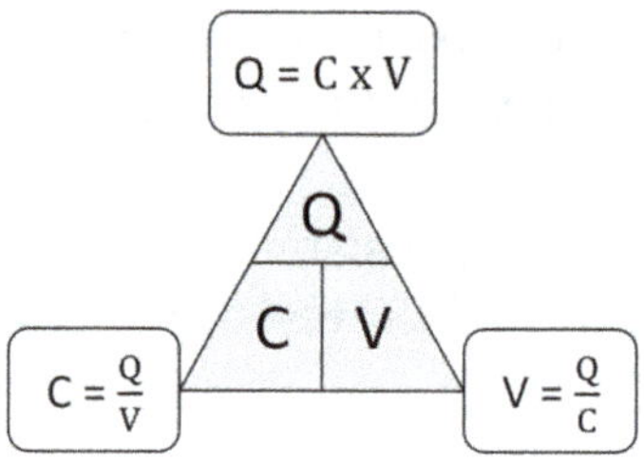

Capacitance, Charge and Voltage of Capacitor

The capacitor symbol looks like two parallel lines with one line is either curve or flat. There are two types circuit symbols of capacitor; Non-polarized and polarized.

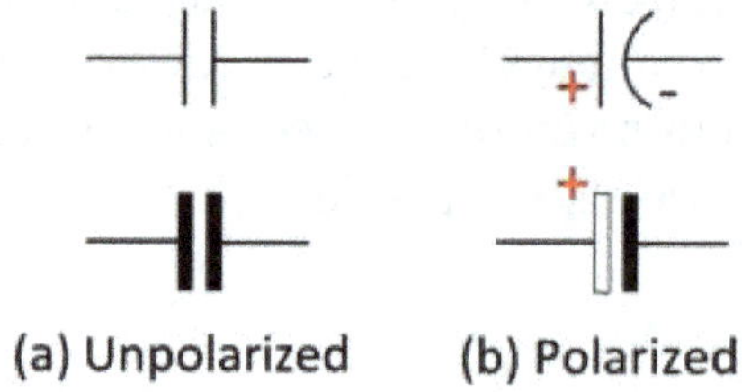

Circuit Symbols of Capacitor

Polarized component is asymmetric. Polarity makes component unidirectional. It gives component a unique position of placement in circuit.

Construction of a Capacitor

The basic structure of capacitor consists of two parallel metal foils (very thin sheets of metal). The metal foils act as electrode. Dielectric material is use as as insulator to separate metal foils. The term "di" in dielectric refers to placement in between two (di) foils and electric means it holds electric field. The circuit symbol of capacitor nearly looks like basic structure of capacitor.

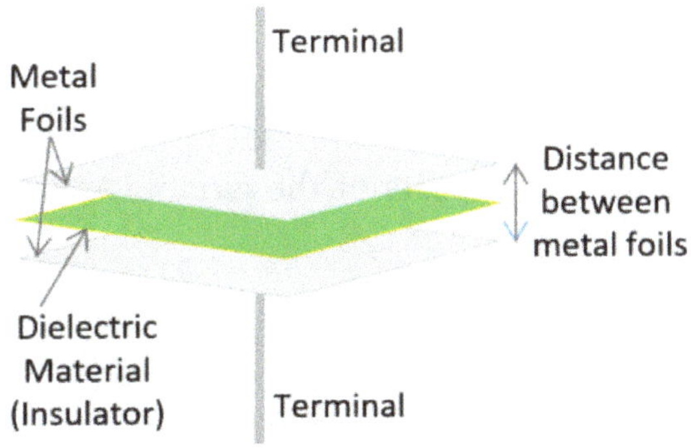

Basic Structure of Capacitor

The insulating material such as glass, rubber, ceramic, plastic, paper etc are use as dielectric material. The material use as metal foils are tantalum, aluminum, mica etc. The operation and structure of capacitor defines its capacitance as per following two relations,

$$C = \frac{Q}{V} = \frac{\epsilon A}{d} \qquad \ldots(\epsilon = \epsilon_0 \, \epsilon_r)$$

Equation of Capacitors Capacitance

Where, ϵ (epsilon) is permittivity (a kind of resistance present when electric field is establish in a medium), ϵ_o is permittivity of air (vacuum) with constant value of 8.85 × 10−12 Farad/meter, ϵ_r is permittivity of dielectric material/relative permittivity/dielectric constant. The typical values of dielectric constant for various commonly use dielectric materials are: Vacuum = 1, Air = 1.006, Paper = 3.85, Mica = 3 – 6, Glass = 3 – 10 etc. Also, 'A' denotes plate area and 'd' is distance between plates.

Working Principle of a Capacitor

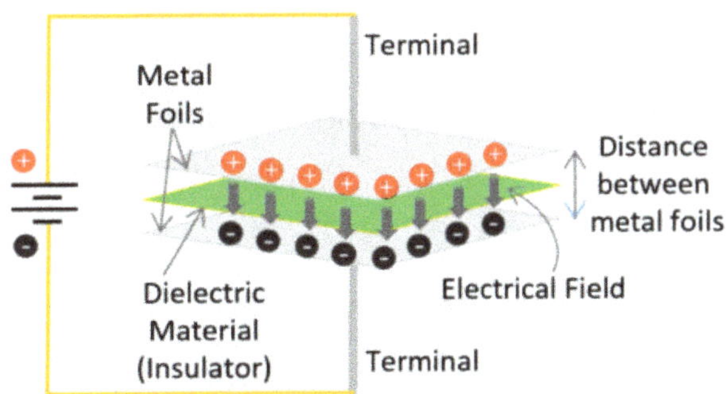

Working of Capacitor

The electric charge is the backbone of component like capacitor. The construction of capacitor shows that there is a gap in between two metal foils. Hence, when electric current passes through capacitor, because of the gap, charge get "freeze" on metal foils. The foil with more electrons get net negative charge and foil with less electrons get positive charge. The dielectric material present in between metal foils do not allow charges to get attracted to each other. Hence these steady charges forms electric field. This structure resembles with the battery. Hence capacitor act as battery. Capacitor and battery stores electrical energy in the form of electrical charge and chemical energy respectively.

Charging and Discharging of Capacitor

Let's take a simple example to understand how capacitance allows capacitor to act like a battery.

There are two switches controlling two parts of the circuit. A capacitor is shown in cylindrical shape and looks like small DC battery.

When upper switch is close and lower switch is open, capacitor is connected with battery and get isolated from other circuit elements. This starts charging of capacitor. Capacitor is kept in charging position for some time, this produces certain amount of voltage inside capacitor.

An LED is present in circuit to demonstrate discharging of capacitor. When upper switch is open and lower switch is close, capacitor get isolated from battery and is connected with other circuit elements. Now capacitor is acting as voltage source. Take a close look at light intensity of LED and voltage drop in capacitor. At start, capacitor easily provides minimum required forward voltage to LED. But when capacitor voltage start to decrease; light intensity of LED also decreases. Finally LED get turn OFF but a small voltage is still remaining in capacitor which is less than minimum forward voltage of LED.

Inductor

The inductor is the simplest of all electronic components, constructed much like a resistor – a simple length of wire that is coiled up. Here, resistance is not the property we're looking for. It is something that happens because of the shape of the wire – a coil – it creates a magnetic field when a current is passed through it. This induced magnetic field gives this bit of wire some interesting electrical properties, especially inductance – which gives these parts their name.

Symbols for Inductors

Like many other electronic components, the symbol for an inductor is a simplified pictogram of what it actually looks like:

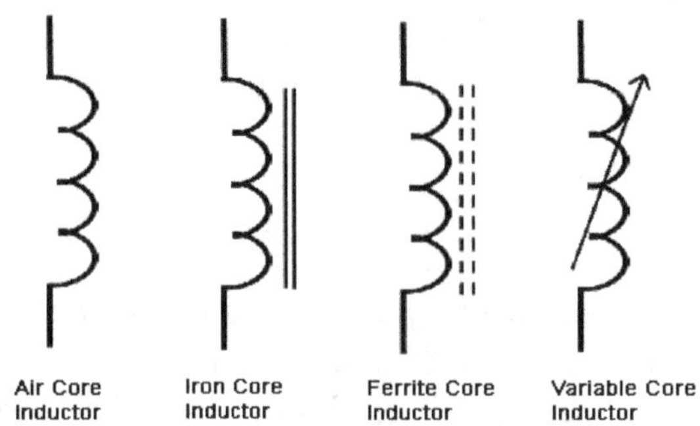

The lines near the symbol represent the core material

Working of an Inductor

An inductor is just a coil of wire.

Any current carrying conductor generates a magnetic field in the following fashion:

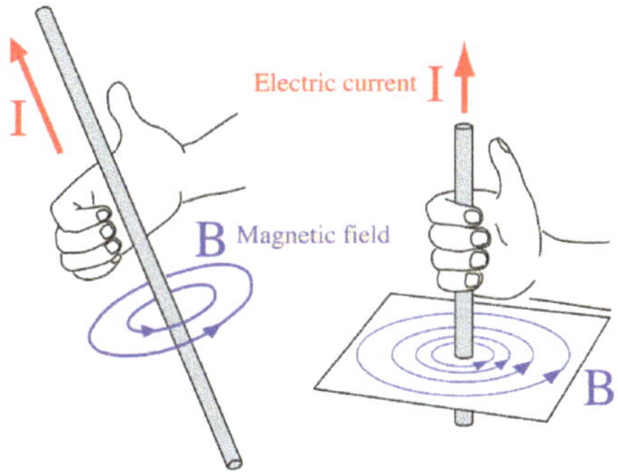

If you plug in the value of the current into the formulas, then you'll realize that the magnetic field produced is tiny – almost negligible, unless the currents are impossibly high, in the order of mega-amps.

So in order to increase the magnetic field created by a specific length of wire, we wrap it into the form of a coil. This increases the magnetic field, like so:

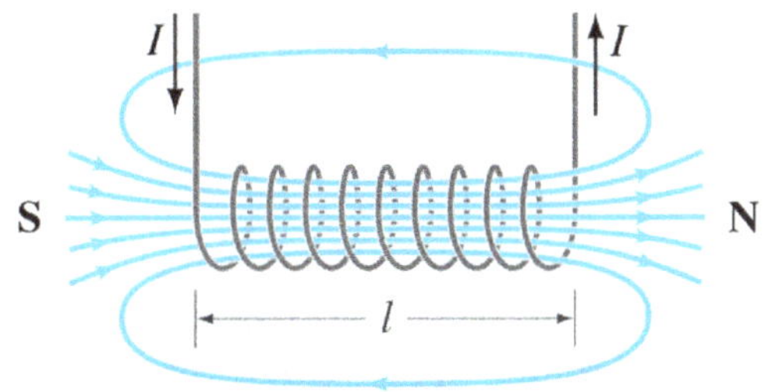

This shape is also called a solenoid.

When a voltage is applied across the terminals of an inductor, the current flowing creates a magnetic field. This magnetic field again creates an induced current in the inductor of opposite polarity, according to Lenz's law. The currents do not cancel each other out – rather, the induced current actively tries to oppose the incoming current due to the voltage across the inductor. The overall result of this battle is that the current through an inductor cannot change rapidly – it is always a linear slope.

Measuring an Inductor

The measuring of inductors can be done by measuring the magnetic field they create. The magnetic field created by an inductor depends on the current that passes through it, so even a small inductor can create a large magnetic field.

Mathematically,

$$V = L(dI/dt)$$

Where V is the voltage, L is the inductance, I is the current and t is the time period.

Inductance, 'L', is measured in Henrys, named after Joseph Henry, the American scientist who discovered electromagnetic induction.

The formula for calculating the inductance of a coil of wire is given by this formula:

L =(μn2a)/l

Where L is the inductance in Henrys, μ is the permeability constant, i.e. a coefficient of how easily the magnetic field can be created in a given medium, n is the number of turns, a is the area of the coil and l is the length of the coil.

The Henry is a very large unit, so practically inductors are measured in microHenrys, uH, which is a millionth of a Henry, or milliHenrys, mH, which is a thousandth of a Henry. Occasionally you might even find very small inductances measured in nanoHenrys, which are a thousandth of a uH.

Different Types of Inductors

Cores are materials that support the creation of a magnetic field. They are generally made of iron and its compounds, such as ferrite (which is an oxide of iron).

1. Air Core Inductors:

Like the name suggests, this kind of inductor has no core – the core material is air! Since air has a relatively low permeability, the inductance of air core inductors is quite low – rarely above 5uH. Since they have a low inductance, the rate of current rise is quite fast for an applied voltage and that makes them capable of handling high frequencies. They are mostly used in RF circuits.

2. Iron Core Inductors

Iron is the most recognizable magnetic material, which makes it an ideal choice for inductors. These take the form of iron-core inductors. They are usually used for low frequency line filtering, since they can be rather beefy and have large inductances. They are also used in audio equipment.

3. Ferrite Core Inductors

Ferrite is just a powder of oxides of iron. This powder is mixed with an epoxy resin and molded to form cores around which wires can be wound. Ferrite core inductors are easily the most recognizable because of their dull grey-black colour. They also are very brittle and break easily. They are the most widely used kinds of inductors, since the permeability can be finely controlled by controlling the ratio of ferrite to epoxy in the mix.

Inductors in Series and Parallel

Inductors connected in series and parallel behave the exact opposite way to capacitors.

For example, to calculate the inductance of a group of inductors in series, you can simply sum up the values of the individual inductances.

$$L = L1 + L2 + \dots + Ln$$

Where L is the total inductance and L1, L2...Ln are the individual inductances.

Suppose you have two inductors, one measuring 10uH and the other 15uH, then by putting them in series you obtain a total inductance of 25uH.

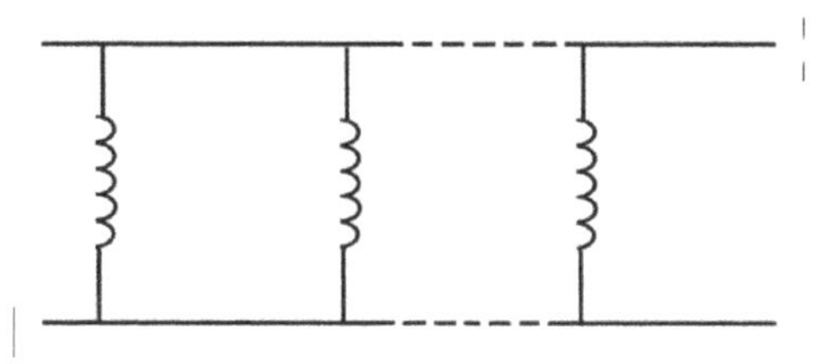

Inductors in parallel behave the same way as resistors in parallel, the inductance is given by:

$$1/L = 1/L1 + 1/L2 + \dots + 1/Ln$$

Where L is the total inductance and L1, L2...Ln are the individual inductances.

In this way if you connect two 10uH inductors in parallel, you'll end up with an inductance of 5uH.

Useful Inductor Formulas

1. Energy Stored By Inductors

Inductors can store energy much like capacitors, but the energy is gone the moment you disconnect the power and the magnetic field collapses. In other words, an unpowered inductor cannot sustain its magnetic field:

$$E = ½ * L * I2$$

Where E is the energy in Joules, L is the inductance in Henries and I is the current in amps.

If you have an inductor of 20uH with a 5A current flowing through it, then the energy stored will be 0.00025J. In this aspect inductors too, like capacitors, hold very little energy.

2. Rate of Current Rise

This formula has already been discussed, but it is worth having a closer look:

$$V/L = dI/dt$$

Where V is the voltage applied across the inductor, L is the inductance, I is the current and t is the time.

This states that when a constant voltage is applied across the inductor, the current rises in a linear slope. This can be useful in creating current ramps, just like the capacitor created voltage ramps at a constant current.

3. Impedance

Inductors have impedance that is related to the frequency by the formula:

$$XL = 2\pi * f * L$$

Where XL is the inductive impedance, f is the frequency in Hertz and L is the inductance in Henrys.

Inductor Behavior in Circuits

Surprisingly, inductors are pretty much useless in DC circuits, since there is a constant current flowing and the inductor acts like a bit of wire.

They find most of their uses in AC circuits. Like mentioned above, they have an impedance, which makes them useful to limit current in an AC circuit, such as fluorescent lamp ballasts.

They can also be used to filter signals.

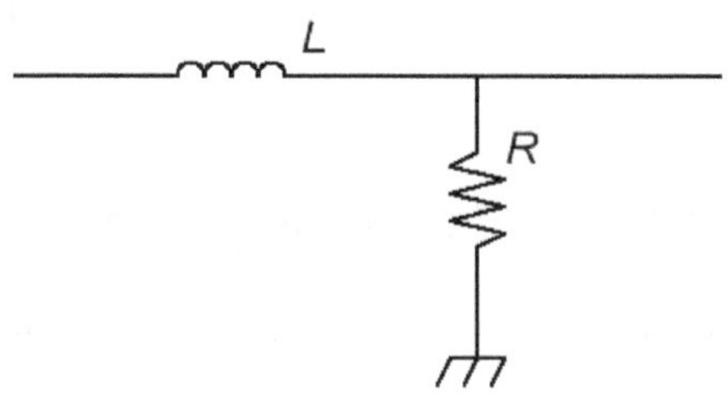

In the first case, the inductor allows all DC current to flow through it to ground, preventing all the low frequencies from reaching the output. At higher frequencies, the impedance of the inductor increases steadily, so the signal can pass through to the output, hence it is called a high pass filter.

In the second case, the inductor allows DC and low frequencies to pass through but blocks all high frequencies from the output and so it is called a low pass filter.

Inductors in Real Life

Inductors, since they are made of copper wire and ferrite, tend to be expensive and find most of their uses in radios, power supplies and telecoms equipment.

In power supplies, the property of an inductor to prevent sudden changes in current is used. Along with a capacitor, it acts to prevent sudden changes in the power supply output voltage and current.

RF circuits make use of an interesting LC circuit called a tank. The capacitor is charged and discharged into the inductor, which builds up its magnetic field. When the magnetic field collapses, a voltage is created and charges the capacitor. This creates periodic oscillations which can be used to generate high frequencies.

The frequency can be calculated by the formula:

$$f = 2\pi\sqrt{LC}$$

Where f is the frequency in Hertz, L is the inductance in Henrys and C is the capacitance in Farads.

Amplifier

An amplifier is an electronic device or circuit which is used to increase the magnitude of the signal applied to its input.

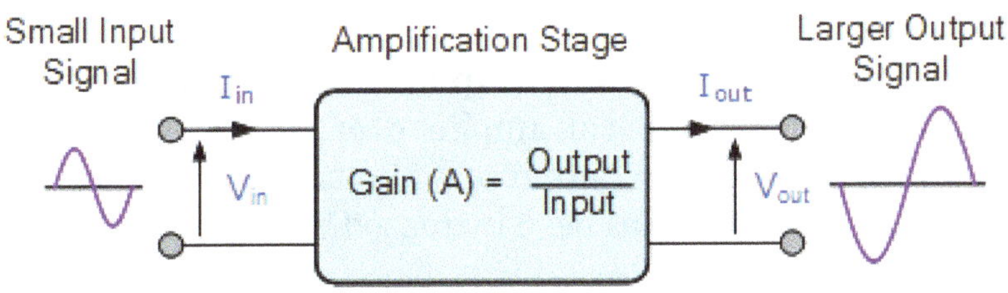

Amplifier is the generic term used to describe a circuit which produces and increased version of its input signal. However, not all amplifier circuits are the same as they are classified according to their circuit configurations and modes of operation.

In "Electronics", small signal amplifiers are commonly used devices as they have the ability to amplify a relatively small input signal, for example from a Sensor such as a photo-device, into a much larger output signal to drive a relay, lamp or loudspeaker for example.

There are many forms of electronic circuits classed as amplifiers, from Operational Amplifiers and Small Signal Amplifiers up to Large Signal and Power Amplifiers. The classification of an amplifier depends upon the size of the signal, large or small, its physical configuration and how it processes the input signal, that is the relationship between input signal and current flowing in the load.

The type or classification of an Amplifier is given in the following table.

Classification of Signal Amplifier

Type of Signal	Type of Configuration	Classification	Frequency of Operation
Small Signal	Common Emitter	Class A Amplifier	Direct Current (DC)
Large Signal	Common Base	Class B Amplifier	Audio Frequencies (AF)
	Common Collector	Class AB Amplifier	Radio Frequencies (RF)
		Class C Amplifier	VHF, UHF and SHF Frequencies

Amplifiers can be thought of as a simple box or block containing the amplifying device, such as a Bipolar Transistor, Field Effect Transistor or Operational Amplifier, which has two input terminals and two output terminals (ground being common) with the output signal being much greater than that of the input signal as it has been "Amplified".

An ideal signal amplifier will have three main properties: Input Resistance or (R_{IN}), Output Resistance or (R_{OUT}) and amplification known commonly as Gain or (A). No matter how complicated an amplifier circuit is, a general amplifier model can still be used to show the relationship of these three properties.

Ideal Amplifier Model

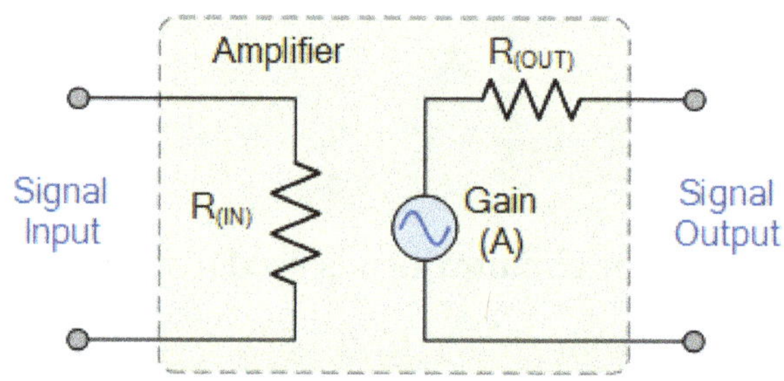

The amplified difference between the input and output signals is known as the Gain of the amplifier. Gain is basically a measure of how much an amplifier "amplifies" the input signal. For example, if we have an input signal of 1 volt and an output of 50 volts, then the gain of the amplifier would be "50". In other words, the input signal has been increased by a factor of 50. This increase is called Gain.

Amplifier gain is simply the ratio of the output divided-by the input. Gain has no units as its a ratio, but in Electronics it is commonly given the symbol "A", for Amplification. Then the gain of an amplifier is simply calculated as the "output signal divided by the input signal".

Amplifier Gain

The introduction to the amplifier gain can be said to be the relationship that exists between the signal measured at the output with the signal measured at the input. There are three different kinds of amplifier gain which can be measured and these are: *Voltage Gain* (Av), *Current Gain* (Ai) and *Power Gain* (Ap) depending upon the quantity being measured with examples of these different types of gains are given below:

Amplifier Gain of the Input Signal

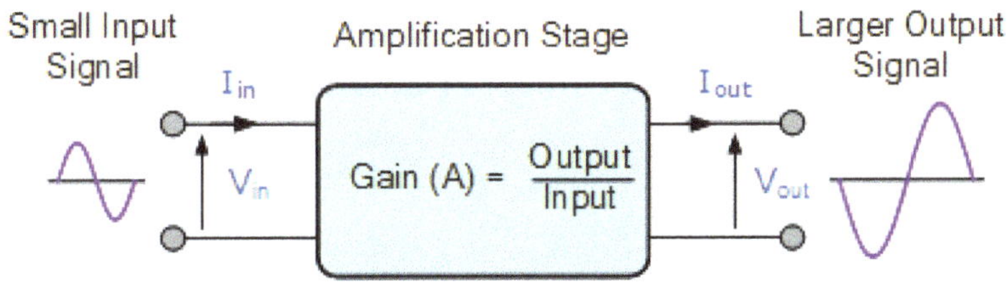

Voltage Amplifier Gain:

$$Current\ Gain\ (A_v) = \frac{\text{Output Current}}{\text{Input Current}} = \frac{Vout}{Vin}$$

Current Amplifier Gain:

$$Current\ Gain\ (A_i) = \frac{\text{Output Current}}{\text{Input Current}} = \frac{Iout}{Iin}$$

Power Amplifier Gain:

$$Power\ Gain(A_p) = A_v\ x\ A_i$$

Note that for the Power Gain you can also divide the power obtained at the output with the power obtained at the input. Also when calculating the gain of an amplifier, the subscripts v, i and p are used to denote the type of signal gain being used.

The power gain (Ap) or power level of the amplifier can also be expressed in Decibels, (dB). The Bel (B) is a logarithmic unit (base 10) of measurement that has no units. Since the Bel is too large a unit of measure, it is prefixed with *deci* making it Decibels instead with one decibel being one tenth (1/10th) of a Bel. To calculate the gain of the amplifier in Decibels or dB, we can use the following expressions.

- Voltage Gain in dB: a_v = 20*log(Av)
- Current Gain in dB: a_i = 20*log(Ai)
- Power Gain in dB: a_p = 10*log(Ap)

Note that the DC power gain of an amplifier is equal to ten times the common log of the output to input ratio, where as voltage and current gains are 20 times the common log of the ratio. 20dB is not twice as much power as 10dB because of the log scale.

Also, a positive value of dB represents a Gain and a negative value of dB represents a Losswithin the amplifier. For example, an amplifier gain of +3dB indicates that the amplifiers output signal has “doubled”, (x2) while an amplifier gain of -3dB indicates that the signal has “halved”, (x0.5) or in other words a loss.

The -3dB point of an amplifier is called the half-power point which is -3dB down from maximum, taking 0dB as the maximum output value.

Amplifier Example No:

Determine the Voltage, Current and Power Gain of an amplifier that has an input signal of 1mA at 10mV and a corresponding output signal of 10mA at 1V. Also, express all three gains in decibels, (dB).

The Various Amplifier Gains:

$$A_V = \frac{\text{Output Voltage}}{\text{Input Voltage}} = \frac{1}{0.01} = 100$$

$$A_i = \frac{\text{Output Current}}{\text{Input Current}} = \frac{10}{1} = 10$$

$$A_P = A_v \times A_i = 100 \times 10 = 1{,}000$$

Amplifier Gains given in Decibels (dB):

$$a_v = 20\log A_v = 20\log 100 = 40\,\text{dB}$$

$$a_i = 20\log A_i = 20\log 10 = 20\,\text{dB}$$

$$a_p = 10\log A_p = 10\log 1000 = 30\,\text{dB}$$

Then the amplifier has a Voltage Gain, (Av) of 100, a Current Gain, (Ai) of 10 and a Power Gain, (Ap) of 1,000.

Generally, amplifiers can be sub-divided into two distinct types depending upon their power or voltage gain. One type is called the Small Signal Amplifier which include pre-amplifiers, instrumentation amplifiers etc. Small signal amplifies are designed to amplify very small signal voltage levels of only a few micro-volts (μV) from sensors or audio signals.

The other type are called Large Signal Amplifiers such as audio power amplifiers or power switching amplifiers. Large signal amplifiers are designed to amplify large input voltage signals or switch heavy load currents as you would find driving loudspeakers.

Power Amplifiers

The Small Signal Amplifier is generally referred to as a "Voltage" amplifier because they usually convert a small input voltage into a much larger output voltage. Sometimes an amplifier circuit is required to drive a motor or feed a loudspeaker and for these types of applications where high switching currents are needed Power Amplifiers are required.

As their name suggests, the main job of a "Power Amplifier" (also known as a large signal amplifier), is to deliver power to the load, and as we know from above, is the product of the voltage and current applied to the load with the output signal power being greater than the input signal power. In other words, a power amplifier amplifies the power of the input signal which is why these types of amplifier circuits are used in audio amplifier output stages to drive loudspeakers.

The power amplifier works on the basic principle of converting the DC power drawn from the power supply into an AC voltage signal delivered to the load. Although the amplification is high the efficiency of the conversion from the DC power supply input to the AC voltage signal output is usually poor.

The perfect or ideal amplifier would give us an efficiency rating of 100% or at least the power "IN" would be equal to the power "OUT". However, in reality this can never happen as some of the power is lost in the form of heat and also, the amplifier itself consumes power during the amplification process. Then the efficiency of an amplifier is given as:

Amplifier Efficiency:

$$Efficiency(\eta) = \frac{\text{power delivered to the load}}{\text{d.c. powor taken from the supply}} = \frac{\text{Pout}}{\text{Pin}}$$

Ideal Amplifier

- The amplifiers gain, (A) should remain constant for varying values of input signal.
- Gain is not be affected by frequency. Signals of all frequencies must be amplified by exactly the same amount.
- The amplifiers gain must not add noise to the output signal. It should remove any noise that is already exists in the input signal.
- The amplifiers gain should not be affected by changes in temperature giving good temperature stability.
- The gain of the amplifier must remain stable over long periods of time.

Electronic Amplifier Classes

The classification of an amplifier as either a voltage or a power amplifier is made by comparing the characteristics of the input and output signals by measuring the amount of time in relation to the input signal that the current flows in the output circuit.

For the transistor to operate within its "Active Region" some form of "Base Biasing" was required. This small Base Bias voltage added to the input signal allowed the transistor to reproduce the full input waveform at its output with no loss of signal.

However, by altering the position of this Base bias voltage, it is possible to operate an amplifier in an amplification mode other than that for full waveform reproduction. With the introduction to the amplifier of a Base bias voltage, different operating ranges and modes of operation can be obtained which are categorized according to their classification. These various mode of operation are better known as Amplifier Class.

Audio power amplifiers are classified in an alphabetical order according to their circuit configurations and mode of operation. Amplifiers are designated by different classes of operation such as class "A", class "B", class "C", class "AB", etc. These different amplifier classes range from a near linear output but with low efficiency to a non-linear output but with a high efficiency.

No one class of operation is "better" or "worse" than any other class with the type of operation being determined by the use of the amplifying circuit. There are typical maximum conversion efficiencies for the various types or class of amplifier, with the most commonly used being:

- Class A Amplifier – has low efficiency of less than 40% but good signal reproduction and linearity.
- Class B Amplifier – is twice as efficient as class A amplifiers with a maximum theoretical efficiency of about 70% because the amplifying device only conducts (and uses power) for half of the input signal.
- Class AB Amplifier – has an efficiency rating between that of Class A and Class B but poorer signal reproduction than Class A amplifiers.
- Class C Amplifier – is the most efficient amplifier class but distortion is very high as only a small portion of the input signal is amplified therefore the output signal bears very little resemblance to the input signal. Class C amplifiers have the worst signal reproduction.

Class A Amplifier Operation

Class A Amplifier operation is where the entire input signal waveform is faithfully reproduced at the amplifiers output as the transistor is perfectly biased within its active region, thereby never reaching either of its cut-off or saturation regions. This then results in the AC input signal being perfectly "centred" between the amplifiers upper and lower signal limits.

Class A Amplifier Output Waveform

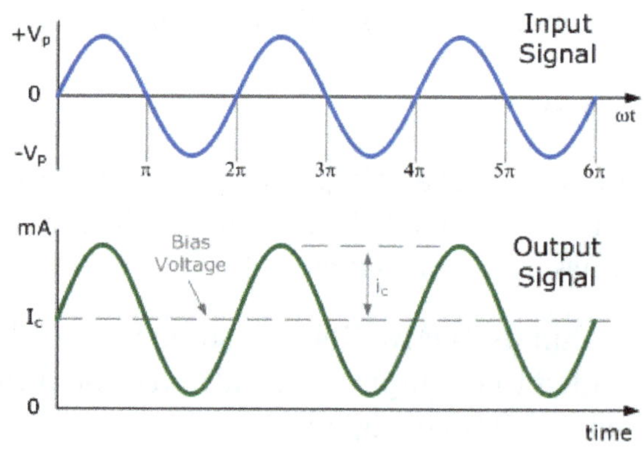

In this configuration, the Class A amplifier uses the same transistor for both halves of the output waveform and due to its biasing arrangement the output transistor always has current flowing through it, even if there is no input signal. In other words the output transistors never turns "OFF". This results in the Class A type of operation being very inefficient as its conversion of the DC supply power to the AC signal power delivered to the load is usually very low.

Generally, the output transistor of a Class A amplifier gets very hot even when there is no input signal present so some form of heat sinking is required. The direct current flowing through the output transistor (Ic) when there is no output signal will be equal to the current flowing through the load. Then a Class A amplifier is very inefficient as most of the DC power is converted to heat.

Class B Amplifier Operation

Unlike the Class A amplifier mode of operation above that uses a single transistor for its output power stage, the Class B Amplifier uses two complimentary transistors (either an NPN and a PNP or a NMOS and a PMOS) for each half of the output waveform. One transistor conducts for one-half of the signal waveform while the other conducts for the other or opposite half of the signal waveform. This means that each transistor spends half of its time in the active region and half its time in the cut-off region thereby amplifying only 50% of the input signal.

Class B operation has no direct DC bias voltage like the class A amplifier, but instead the transistor only conducts when the input signal is greater than the base-emitter voltage and for silicon devices is about 0.7v. Therefore, at zero input there is zero output. This then results in only half the input signal being presented at the amplifiers output giving a greater amount of amplifier efficiency.

Class B Amplifier Output Waveform

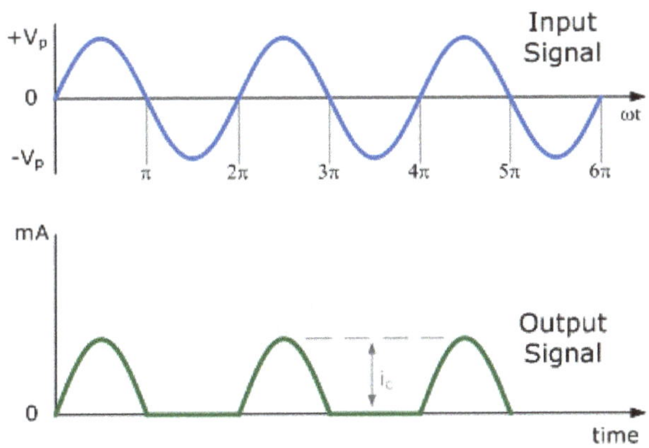

In a Class B amplifier, no DC voltage is used to bias the transistors, so for the output transistors to start to conduct each half of the waveform, both positive and negative, they need the base-emitter voltage Vbe to be greater than the 0.7v required for a bipolar transistor to start conducting.

Then the lower part of the output waveform which is below this 0.7v window will not be reproduced accurately resulting in a distorted area of the output waveform as one transistor turns "OFF" waiting for the other to turn back "ON". The result is that there is a small part of the output waveform at the zero voltage cross over point which will be distorted. This type of distortion is called Crossover Distortion.

Class AB Amplifier Operation

The Class AB Amplifier is a compromise between the Class A and the Class B configurations above. While Class AB operation still uses two complementary transistors in its output stage a very small biasing voltage is applied to the Base of the transistor to bias it close to the Cut-off region when no input signal is present.

An input signal will cause the transistor to operate as normal in its Active region thereby eliminating any crossover distortion which is present in class B configurations. A small Collector current will flow when there is no input signal but it is much less than that for the Class A amplifier configuration.

This means then that the transistor will be "ON" for more than half a cycle of the waveform. This type of amplifier configuration improves both the efficiency and linearity of the amplifier circuit compared to a pure Class A configuration.

Class AB Amplifier Output Waveform

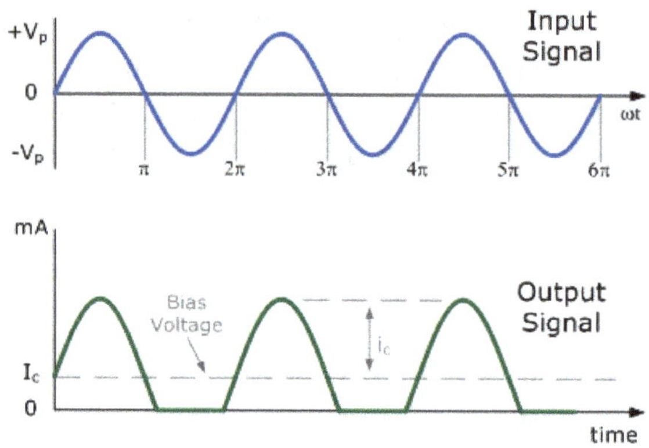

The class of operation for an amplifier is very important and is based on the amount of transistor bias required for operation as well as the amplitude required for the input signal. Amplifier classification takes into account the portion of the input signal in which the transistor conducts as well as determining both the efficiency and the amount of power that the switching transistor both consumes and dissipates in the form of wasted heat. Then we can make a comparison between the most common types of amplifier classifications in the following table.

Power Amplifier Classes

Class	A	B	C	AB
Conduction Angle	360°	180°	Less than 90°	180 to 360°
Position of the Q-point	Centre Point of the Load Line	Exactly on the X-axis	Below the X-axis	In between the X-axis and the Centre Load Line
Overall Efficiency	Poor 25 to 30%	Better 70 to 80%	Higher than 80%	Better than A but less than B 50 to 70%
Signal Distortion	None if Correctly Biased	At the X-axis Crossover Point	Large Amounts	Small Amounts

Badly designed amplifiers especially the Class "A" types may also require larger power transistors, more expensive heat sinks, cooling fans, or even an increase in the size of the power supply required to deliver the extra wasted power required by the amplifier. Power converted into heat from transistors, resistors or any other component for that matter, makes any electronic circuit inefficient and will result in the premature failure of the device.

Diode

A Diode is the simplest two-terminal unilateral semiconductor device. It allows current to flow only in one direction and blocks the current that flows in the opposite direction. The two terminals of the diode are called as anode and cathode. The symbol of diode is as shown in the figure below:

Figure: Symbol of a Diode

The characteristics of a diode closely match to that of a switch. An ideal switch when open does not conduct current in either directions and in closed state conducts in both directions. The characteristic of a diode is as shown in the figure below:

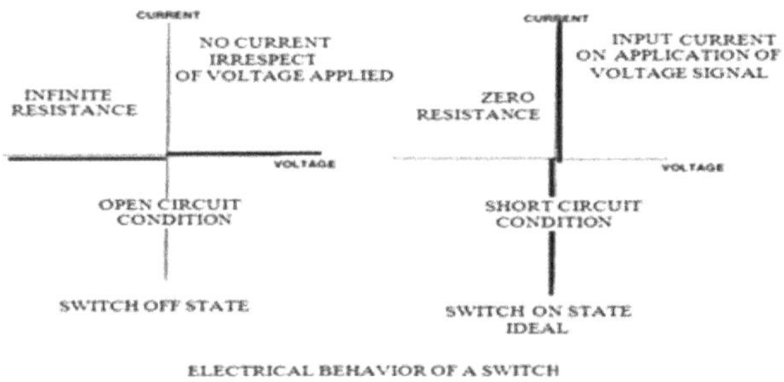

Figure: Graph showing electrical characterstics of an Ideal Diode

Ideally, in one direction that is indicated by the arrow head diode must behave short circuited and in other one that opposite to that of the direction of arrow head must be open circuited. By ideal characteristics, the diodes is designed to meet these features theoretically but are not achieved practically. So the practical diode characteristics are only close to that of the desired.

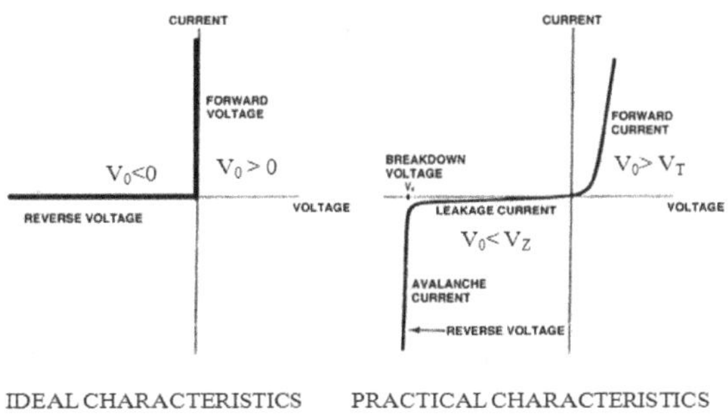

Figure: Graph showing comparison of electrical characterstics of Ideal and Practical Diodes

Working Principle of Diode

The diode operates when a voltage signal is applied across its terminals. The application of a DC voltage to make the diode operate in a circuit is called as 'Biasing'. As already mentioned above the diode resembles to that of a one way switch so it can either be in a state of conduction or in a state of non conduction. The 'ON' state of a diode is achieved by 'Forward biasing' which means that positive or higher potential is applied to the anode and negative or lower potential is applied at the cathode of the diode. In other words, the 'ON' state of diode has the applied current in the same direction of the arrow head. The 'OFF' state of a diode is achieved by 'Reverse biasing' which means that positive or higher potential is applied to the cathode and negative or lower potential is applied at the anode of the diode. In other words, the 'OFF' state of diode has the applied current in the opposite direction of the arrow head.

During 'ON' state, the practical diode offers a resistance called as the 'Forward resistance'. The diode requires a forward bias voltage to switch to the 'ON' condition which is called Cut-in-voltage. The diode starts conducting in reverse biased mode when the reverse bias voltage exceeds its limit which is called as the Breakdown voltage. The diode remains in 'OFF' state when no voltage is applied across it.

A simple p-n juction diode is fabricated by doping p and n type layers on a silicon or germanium wafer. The germanium and silicon materials are prefered for diode fabrication because:

- They are available in high purity.
- Slight doping like one atom per ten million atoms of a desired impurity can change the conductivity to a considerable level.
- The properties of these materials change on applying heat and light and hence it is important in the devlopment of heat and light sensetive devices.

Types of Diodes

The other variant of diodes have different construction, characteristics and applications. The different types of diodes are:

- Small signal or Small current diode - These diodes assumes that the operating point is not affected because the signal is small.
- Large signal diodes - The operating point in these diodes get affected as the signal is large.
- Zener diodes - This diode runs in reverse bias condition when the voltage reaches the breakdown point. A stable voltage can be achieved by placing a resistor across it to liimit the current. This diode is used to provide reference voltage in power supply circuits.
- Light emitting diodes (LED) - This is the most popular kind of diode. When it works in the forward bias condition, the current flows through the junction to produce the light.
- Photodiodes - The electrons and holes are generated as light strikes across the p-n junction causing the current to flow. Theses diodes can work as photodetector and are used to generate electricity.
- Constant current diodes - This diode keeps the current constant even when the voltage applied keeps changing. It consists of JFET (junction – field effect transistor) with the source shorted to the gate in order to function like a two - terminal current limiter or current source.

- Schottky diode - These diodes are used in RF applications and clamping circuits. This diode has lower forward voltage drop as against the silicon PN junction diodes.
- Shockley diode - This is a four layer diode which is also known as PNPN diode. This didoe is similar to thyristor where the gate is disconnected.
- Step recovery diodes - This semiconductor diode has the ability to generate short pulses and hence it is used in microwave applications as a pulse generator.
- Tunnel diodes - This diode is heavily doped in the forward bias condition that has a negative resistance at extremely low voltage and a short circuit in the negative bias direction. This diode is useful as a microwave ampilifer and in oscillators.
- Varactor diodes - This diode works in reverse bias condition and restricts the flow of current thorugh the junction. Depending on the amount of biasing, the width of the depletion region keeps varying. This diode comprises of two plates of a capacitor with the depletion region amidst them. The variation in capacitance depends upon the depletion region and this can varied by altering the reverse bias on the diode.
- PIN diodes - This diode has intrinsic semiconductor sandwiched between P- type and N- type region. Doping does not occur in this type of diode and thereby the intrinsic semiconductor increases the width of the depletion region. They are used as ohtodiodes and radio frequency switches.
- LASER diode - This diode produces laser type of light and are expensive as compared to LED. They are widely used in CD and DVD drives.
- Transient voltage supression diodes - This diode is used to protect the electronics that are sensitive against voltage spikes.
- Gold doped diodes - These diodes use gold as the dopant and can operate at signal frequencies even if the forward voltage drop increases.
- Super barrier diodes - These are also called as the rectifier diodes. This diodes have the property of low reverse leakage current as that of normal p-n junction diode and low forward voltage drop as that of Schottky diode with surge handling ability.
- Point contact diodes - The construction of this diode is simpler and are used in analog applications and as a detector in radio receivers. This diode is built of n – type semiconductor and few conducting metals placed to be in contact with the semiconductor. Some metals move from towards the semiconductor to form small region of p- tpye semiconductor near the contact.
- Peltier diodes - This diode is used as heat engine and sensor for thermoelectric cooling.
- Gunn diode - This diode is made of materials like GaAs or InP that exhibit a negative differential resistance region.
- Crystal diode - These are a type of point contact diodes which are also called as Cat's whisker diode. This didoe comprises of a thin sharpened metal wire which is pressed against the

semiconducting crystal. The metal wire is the anode and the semconducting crystal is the cathode. These diodes are obsolete.

- Avalanche diode - This diode conducts in reverse bias condition where the reverse bias volage applied across the p-n junction creates a wave of ionization leading to the flow of large current. These didoes are designed to breakdown at specific reverse voltage in order to avoid any damage.
- Silicon controlled rectifier - As the name implies this diode can be controlled or triggered to the ON condition due to the application of small voltage. They belong to the family of Thyristors and is used in various fields of DC motor control, generator field regulation, lighting system control and variable frequency drive. This is three terminal device with anode, cathode and third controled lead or gate.
- Vaccum diodes - This diode is two electrode vacuum tube which can tolerate high inverse voltages.

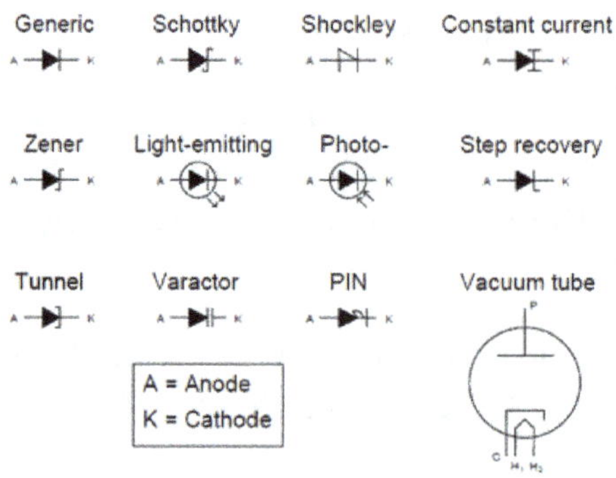

various symbols of Diode

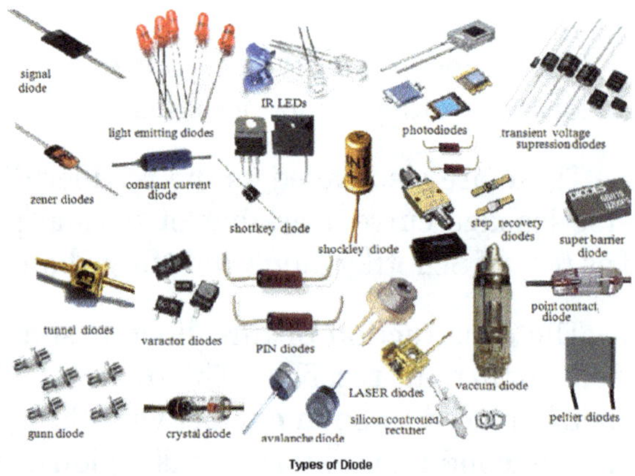

Figure: Image showing various types of diodes

Generic Diodes (Small Signal and Large Signal)

A p-n junction diode is the simplest semiconductor device. It is a two-terminal, bipolar, unilateral rectifying device that conducts only in one direction. The generic diodes are used in the following fields:

- Rectification in power supply circuits.

- Extraction of modulation from radio signals in a radio receiver and in protection circuits where large transient currents may appear on low current transistors or ICs in interfacing with relays or other high power devices.

Used in series with power inputs to electronic circuits where only one of negative or positive polarity voltage is desired.

Construction

A simple p-n diode is a junction where p-type and n-type layers are doped on a silicon or germanium wafer. A p-type semiconductor is formed by doping of trivalent or acceptor impurity atoms on a pure silicon or germanium thereby having an excess concentration of holes. An n-type semiconductor is formed by doping of pentavalent or donor impurity atoms on a pure silicon or germanium thereby having an excess concentration of electrons. So, holes are the majority charge carriers in a p-type region whereas electrons in the n-type region. Electron-holes pairs are thermally generated in both types which constitute the minority charge carriers. It is remarkable that a p-type material is not positively charged in spite of having excessive holes while an n-type material is not negatively charged in spite of excessive electrons. This is because in a p-type material along with holes, the anions are generated and the total number of protons and electrons still remain the same. This is similarly observed for the n-type material.

The junction of a p-type and n-type doping on silicon or germanium wafer produces a small region of the order of micrometers which is depleted from the free charge carriers. This region is formed due to diffusion of holes from a p-type and electrons from an n-type material called as the depletion region or space charge region or transition region. The p-type region to the left of the depletion region is having acceptor negative ion layer and to the right are donor positive ion layer which induces an electric flux or potential difference across the junction. The charge concentration is positive on left of the junction and negative on the right of the junction. This potential barrier stops the holes to migrate into n-type region and electrons to migrate into p-type region as the potential rises for holes and electrons will allow migrating in to n-type and p-type regions. The charge carrier regions around the depletion regions are also called as the uncovered regions.

It is also important that the minority charge currents i.e. electron current in p-type region and hole current in n-type region decreases exponentially across the diode length. The minority current is due to electron hole pairs generated thermally and dependent upon temperature. These currents are so small in magnitude in the order of microamperes. However in conduction state, the current through the diode crystal remains stable. The total current is a sum of minority and majority charge currents due to bipolar nature of the diode. The majority charge currents is hole current in p-type and electron current in n-type are reduced as they migrate near junction due recombination. The minority currents is electron current in p-type and hole current in n-type are maximum near junction and reduces as they migrate away from the junction as an exponential function. The majority charge currents in their regions after crossing the junction are the diffusion currents while before junction are drift currents.

Concept of Ohmic contacts – In addition to PN junction diode, there is a two metal semiconductor junctions originating from the leads in order to connect the device. It is assumed that the resistance of these metal semiconductor contacts remain constant despite of the magnitude and direction of current. During the diode operation, the applied voltage is solely effective for increasing or decreasing the potential barrier height of the PN junction.

Principle and Operation

The possible configurations for a diode are:

1. Open circuited
2. Short circuited
3. Forward biased
4. Reverse biased

1. Open circuited: In open circuited condition, the current that flows through the diode is zero (I = 0). The potential barrier at the PN junction remains the same as created in the diode fabrication.

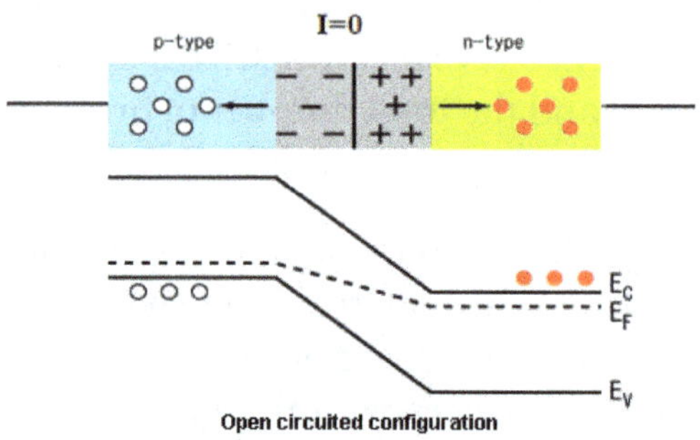

Figure: Image showing working of diode in open circuit configuration

2. Short circuited: In short circuited condition, the sum voltage in the loop must be zero. So it is assumed, that the potential barrier at the PN junction is compensated by the potential drops at the metal semiconductor junctions. The holes supplied by the n-region must be driven to the p region which is physically impossible. The similar discussion applies to the electron current in the n-region.

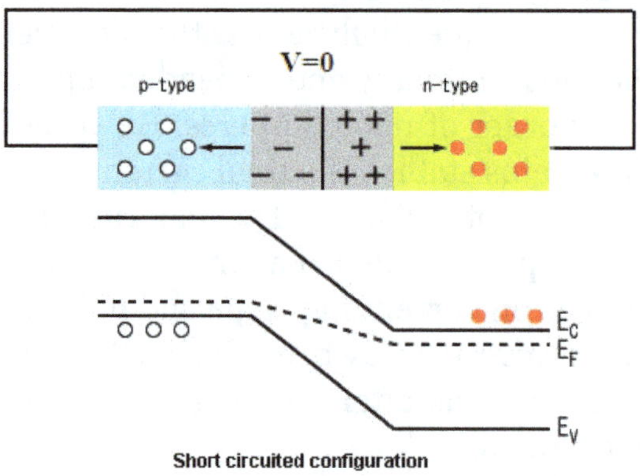

Figure: Image showing working of diode in short circuit configuration

3. Forward bias: In forward bias condition, higher or positive potential is applied at the anode and negative or lower potential is applied at the cathode of a diode. The positive potential at anode

repels the holes in p-region towards n-region while negative potential at the cathode repels electrons in n-region towards p-region. Thus, the height of the potential barrier reduces. The depletion region disappears when the applied voltage equals to the potential barrier and a large current flows through the diode. The voltage required to drive the diode into a state of conduction is called as the 'Cut in/Offset/Threshold/Firing voltage'. The current is of considerable magnitude as it is dominantly constituted by the majority charge currents that is the hole current in the p-region and the electron current in the n-region. The current that flows from anode to cathode is limited by the crystal bulk resistance, recombination of charges and the ohmic contact resistances at the two metal semiconductor junctions. The current is restricted to milli Amperes order.

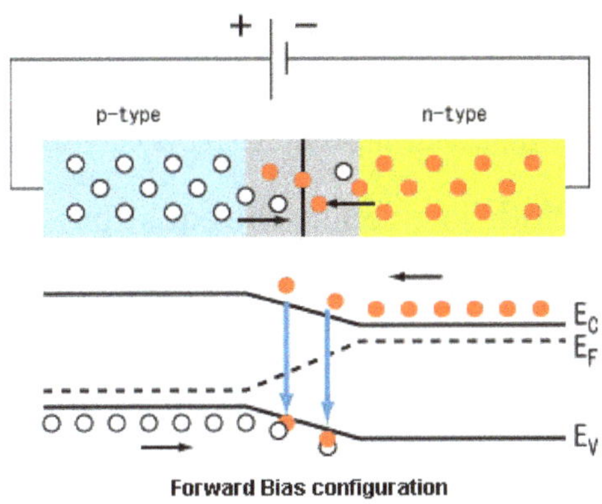

Figure: Image showing working of diode in forward bias configuration

4. Reverse Bias: In reverse bias condition, the higher or positive potential is applied at the cathode and negative or lower potential is applied at the anode. The negative potential at anode attracts the holes in p-region that are away from the n-region while positive potential at the cathode attracts electrons in n-region that are away from the p-region. The applied voltage increases the height of the potential barrier. The current flows dominantly due to the minority charge currents that is the electron current in p-region and the hole current in n-region. Thus a constant current of negligible magnitude flows in the reverse direction which is called as the 'Reverse saturation current'. Practically, the diode remains in a state of non – conduction. The reverse saturation current is of the order of microamperes in a germanium diode or nanoamperes in a silicon diode If the reverse voltage exceeds the limit of 'breakdown/Zener/Peak inverse/Peak reverse voltage', the potential breakdown that occurs leads to a large reverse current.

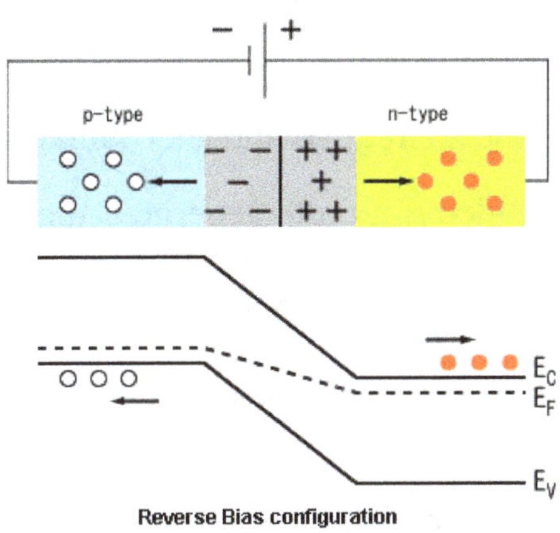

Figure: Image showing working of diode in reverse bias configuration

Characteristics

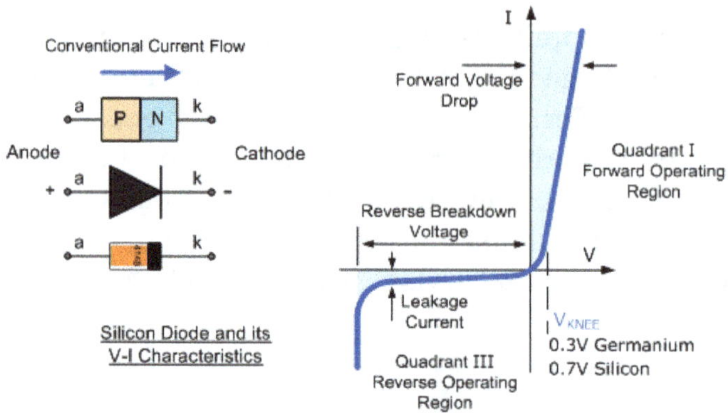

Figure: Graph showing characterstics curve of diode

The current that flows through a diode is given by the equation:

$$I_D = IS\left(e^{V/\eta VT} - 1\right),$$

where I_D - diode current. (Positive for forward and negative for reverse)

I_S - constant reverse saturation current

V - applied voltage. (Positive for forward and negative for reverse)

η - factor dependent upon the nature of semiconductor. (1 for germanium and 2 for silicon)

V_T - volt equivalent of temperature which is given by T/11600. (T is Temperature in Kelvin)

When a forward voltage is applied at the terminals of a diode, the diode begins to conduct. During conduction, the cut in or threshold voltage exceeds the applied forward voltage. The threshold voltage for a germanium diode is 0.3V and for silicon diode is 0.7V. The forward current (miliampere range) initially increases linearly and then increases exponentially for high currents.

When a a reverse voltage is applied, a reverse saturation current flows through the diode. The diode continues to be in the non conducting state until the reverse voltage drops below the zener voltage. As the reverse voltage approximates the peak inverse voltage a breakdown called as thc 'Avalanche breakdown' occurs. During the breakdown, the minority charge carriers ionize the stable atoms which are followed by a chain ionization to generate a large number of free charge carriers. Thus the diode becomes short circuited and gets damaged.

As the temperature increases, the electron pairs generated thermally also increases thereby increasing the conductivity in both directions. The reverse saturation current also increases with the increase in temperature. The change is 11% per °C for a germanium diode and 8% per °C for a silicon diode. On the other hand the diode current is doubled for every 10 °C rise. With increase in voltage, the firing voltage in forward characteristics is reduced while peak reverse voltage is increased.

The peak inverse voltage can be reduced by increasing the doping level. The same concept is used to design zener diodes.

1. Diode resistances: The resistance associated with the diode can be evaluated in three fashions and the three types of resistances associate with a diode accordingly.

- DC or Static resistance: It is the ratio of diode voltage to the diode current at any point of its characteristic curves. It is defined at a point on the characteristic curves.
- AC or dynamic resistance: It is the ratio of change in diode voltage to the change in diode current. It is defined at a point on the characteristic curves over a tangent.
- Average AC resistance: It is the ratio of change in diode voltage to the change in diode current over a straight line joining two limits of operation.

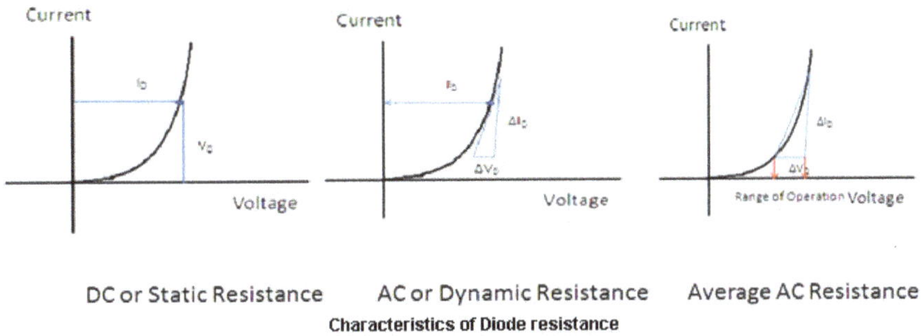

Figure: Graph showing characterstics of diode resistance

2. Diode capacitances: The diode exhibits two types of capacitances transition capacitance and diffusion capacitance.

- Transition capacitance: The capacitance which appears between positive ion layer in n-region and negative ion layer in p-region.
- Diffusion capacitance: This capacitance originates due to diffusion of charge carriers in the opposite regions.

The transition capacitance is very small as compared to the diffusion capacitance.

In reverse bias transition, the capacitance is the dominant and is given by:

$$C_T = \in A / W$$

where, C_T - transition capacitance

A - diode cross sectional area

W - depletion region width

In forward bias, the diffusion capacitance is the dominant and is given by:

$$C_D = dQ/dV = \tau * dI/dV = \tau * g = \tau / r(\text{general})$$

where, C_D - diffusion capacitance

dQ - change in charge stored in depletion region

V - change in applied voltage

τ - time interval for change in voltage

g - diode conductance

r - diode resistance

The diffusion capacitance at low frequencies is given by the formula:

$$C_D = \tau * g / 2 (\text{low frequency})$$

The diffusion capacitance at high frequencies is inversely proportional to the frequency and is given by the formula:

$$C_D = g(\tau/2\omega)^{1/2}$$

Diode switching time: In AC applications, when diode is instantaneously switched from a conduction state to a non conduction state it needs some time to return to non conduction state and behaves short circuited for a little time period in reverse direction. This occurs because when the diode biasing is suddenly changed, the majority charge carriers migrated to other region is the minority charge carriers in the region. Specifically, holes are the minority carriers migrated from p-type to n-type in reverse bias.. These holes require some time to return back to state of non conduction which is called as the 'Reverse recovery time'. Reverse recovery time is the sum of storage time and the transition time.

- Storage time: The time period for which diode remains in conduction state even in the reverse direction.
- Transition time: The time elapsed in returning back to state of non conduction.

It is desirable those diodes has minimum switching or reverse recovery time t_{rr}. Switching time of diodes is of the order of few nanoseconds to 1 microsecond. Now fast switching diodes with switching time up to few picoseconds are also available.

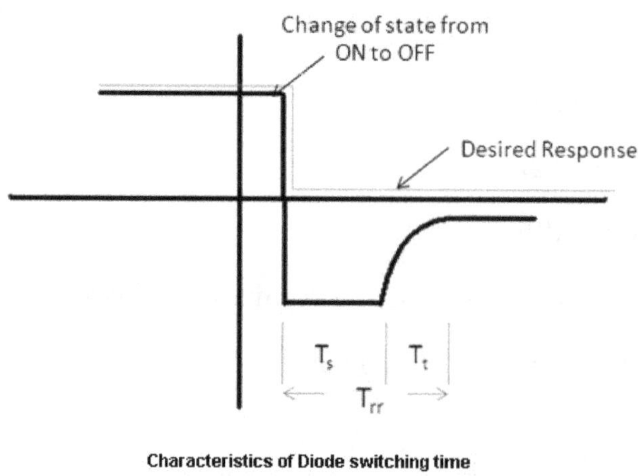

Figure: Graph showing characterstics of diode switching time

Identification

A diode is marked with a bar which indicates the cathode terminal of a diode which is as shown in the figure below:

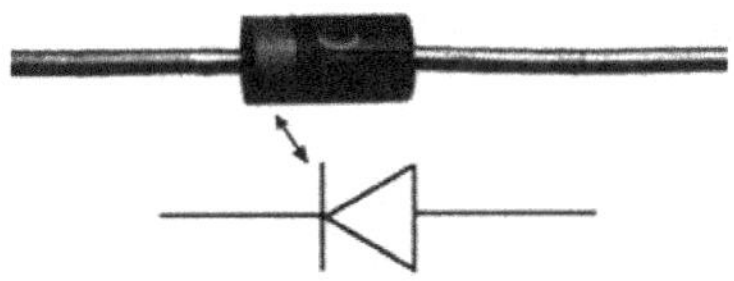

Figure: Image showing identification of diode terminals

Applications

Diodes are used in various applications like rectification, clipper, clamper, voltage multiplier, comparator, sampling gates and filters.

1. Rectification – The rectification means converting AC voltage into DC voltage. The common rectification circuits are half wave rectifier (HWR), full wave rectifier (FWR) and bridge rectifier.

- Half wave rectifier: This circuit rectifies either positive or negative pulse of the input AC. The figure is as shown below:

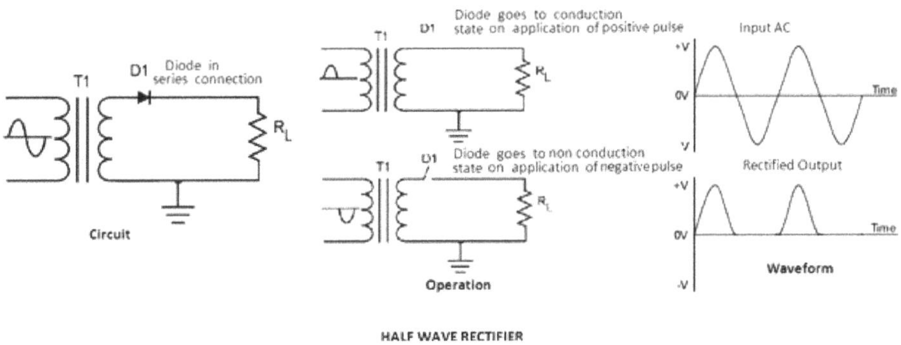

Figure: Circuit Diagram of Diode based Half Wave Rectifier

- Full wave rectifier: This circuit converts the entire AC signal into DC. The figure is as shown below:

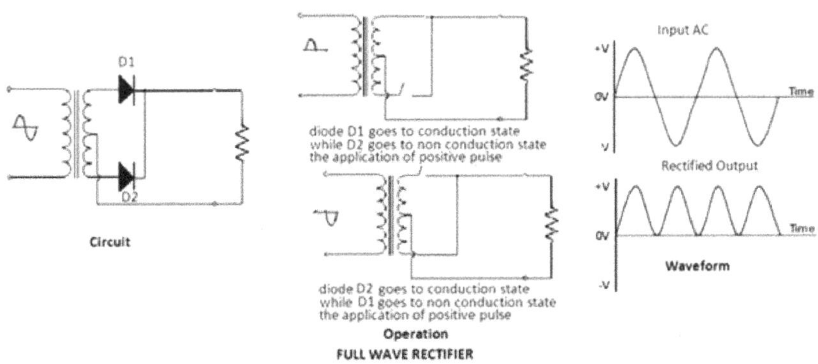

Figure: Circuit Diagram of Diode based Full Wave Rectifier

- Bridge rectifier: This circuit converts the entire AC signal into DC. The figure is as shown below:

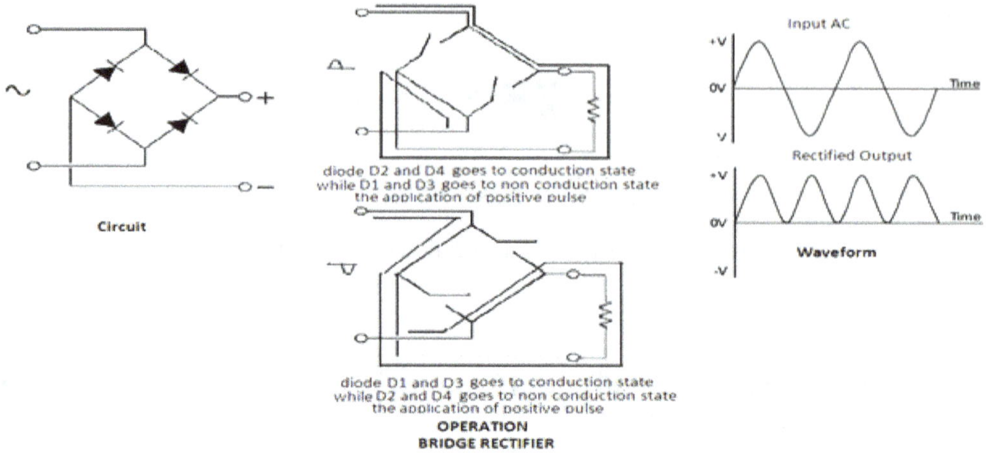

Figure: Circuit Diagram of Diode based Bridge Rectifier

2. Clipper- Diode can be used to clip off some portion of pulse without distorting the remaining part of the waveform. The figure is as shown below:

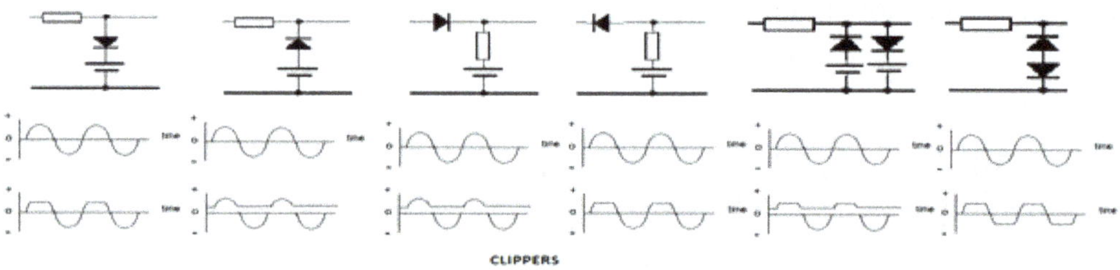

Figure: Circuit Diagram of Diode based Clipper

3. Clamper – A clamping circuit restricts the voltage levels to exceed a limit by shifting the DC level. The peak to peak is not affected by clamping. Diodes with resistors and capacitors are used to make clamping circuits. Sometimes independent DC sources can be used to provide additional shift. The figure is as shown below:

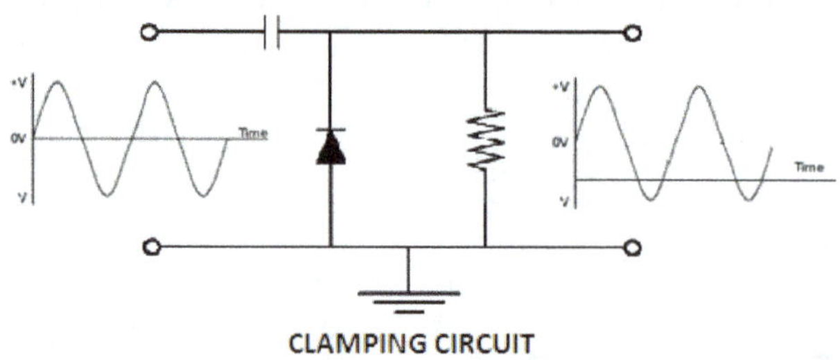

Figure: Circuit Diagram of Diode based Clamper

References

- Electronic-components: electronicscoach.com, Retrieved 4 April, 2019
- Resistor: whatis.techtarget.com, Retrieved 1 August, 2019

- Resistor, electric: rapidtables.com, Retrieved 23 June, 2019
- Resistor: electronics-tutorials.ws, Retrieved 19 March, 2019
- Capacitor, electric: rapidtables.com, Retrieved 13 January, 2019
- What-is-capacitor: binaryupdates.com, Retrieved 9 July, 2019
- Introduction-to-inductors: components101.com, Retrieved 18 March, 2019
- Amplifier: electronics-tutorials.ws, Retrieved 6 May, 2019
- Diodes, tutorials: engineersgarage.com, Retrieved 25 February, 2019

Transistors: A Comprehensive Study

A semiconductor device that is used to amplify signals and electronic powers is termed as a transistor. It is made up of semiconductor material with at least three terminals for connection to an external circuit. Field-effect transistor and bipolar junction transistor are the two main types of transistors. The diverse applications of transistors in the current scenario have been thoroughly discussed in this chapter.

The transistor is a semiconductor device which transfers a weak signal from low resistance circuit to high resistance circuit. The words trans mean transfer property and istor mean resistance property offered to the junctions. In other words, it is a switching device which regulates and amplify the electrical signal likes voltage or current.

The transistor consists two PN diode connected back to back. It has three terminals namely emitter, base and collector. The base is the middle section which is made up of thin layers. The right part of the diode is called emitter diode and the left part is called collector-base diode. These names are given as per the common terminal of the transistor. The emitter based junction of the transistor is connected to forward biased and the collector-base junction is connected in reverse bias which offers a high resistance.

Transistor Symbols

There are two types of transistor, namely NPN transistor and PNP transistor. The transistor which has two blocks of n-type semiconductor material and one block of P-type semiconductor material is known as NPN transistor. Similarly, if the material has one layer of N-type material and two layers of P-type material then it is called PNP transistor. The symbol of NPN and PNP is shown in the figure below:

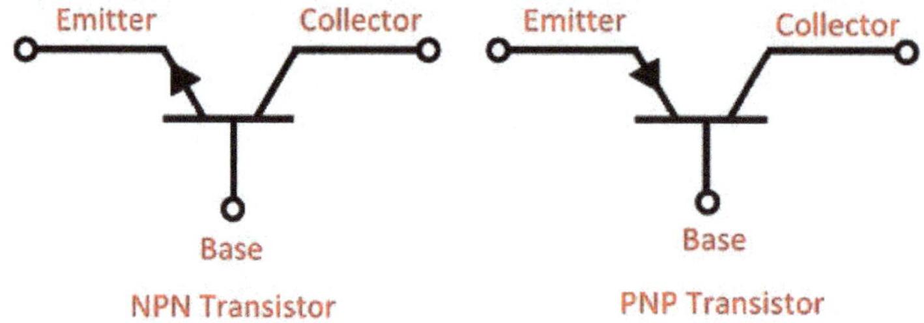

The arrow in the symbol indicates the direction of flow of conventional current in the emitter with forward biasing applied to the emitter-base junction. The only difference between the NPN and PNP transistor is in the direction of the current.

Transistor Terminals

The transistor has three terminals namely, emitter, collector and base. The terminals of the diode are explained below in details.

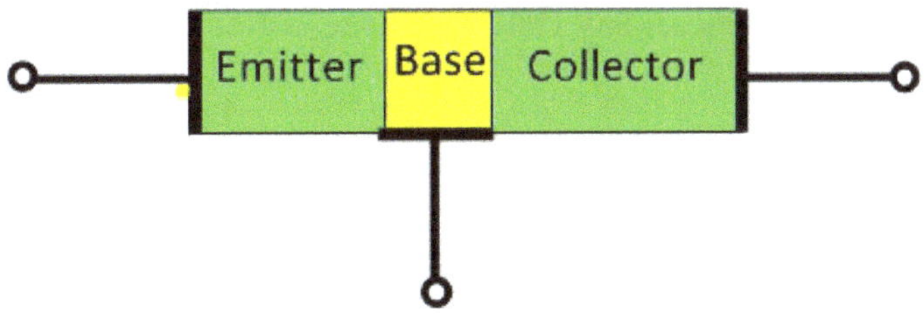

- Emitter – The section that supplies the large section of majority charge carrier is called emitter. The emitter is alway connected in forward biased with respect to the base so that it supplies the majority charge carrier to the base. The emitter-base junction injects a large amount of majority charge carrier into the base because it is heavily doped and moderate in size.
- Collector – The section which collects the major portion of the majority charge carrier supplied by the emitter is called a collector. The collector-base junction is always in reverse bias. Its main function is to remove the majority charges from its junction with the base. The collector section of the transistor is moderately doped, but larger in size so that it can collect most of the charge carrier supplied by the emitter.
- Base – The middle section of the transistor is known as the base. The base forms two circuits, the input circuit with the emitter and the output circuit with the collector. The emitter-base circuit is in forward biased and offered the low resistance to the circuit. The collector-base junction is in reverse bias and offers the higher resistance to the circuit. The base of the transistor is lightly doped and very thin due to which it offers the majority charge carrier to the base.

Working of Transistor

Usually, silicon is used for making the transistor because of their high voltage rating, greater current and less temperature sensitivity. The emitter-base section kept in forward biased constitutes the base current which flows through the base region. The magnitude of the base current is very small. The base current causes the electrons to move into the collector region or create a hole in the base region.

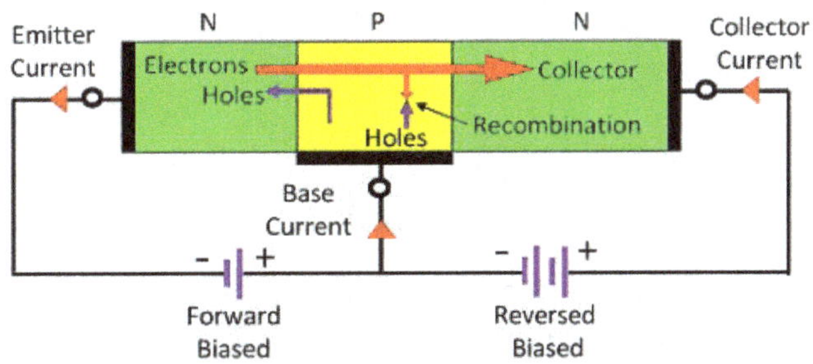

Working of Transistor

The base of the transistor is very thin and lightly doped because of which it has less number of electrons as compared to the emitter. The few electrons of the emitter are combined with the hole of the base region and the remaining electrons are moved towards the collector region and constitute the collector current. Thus we can say that the large collector current is obtained by varying the base region.

Transistor Operating Conditions

When the emitter junction is in forward biased and the collector junction is in reverse bias, then it is said to be in the active region. The transistor has two junctions which can be biased in different ways. The different working conduction of the transistor is shown in the table.

Condition	Emitter Junction (EB)	Collector Junction (CB)	Region of Operation
FR	Forward-biased	Reversed-biased	Active
FF	Forward-biased	Forward-biased	Saturation
RR	Reversed-biased	Reversed-biased	Cut-off
RF	Reversed-biased	Forward-biased	Inverted

1. FR – In this case, the emitter-base junction is connected in forward biased and the collector-base junction is connected in reverse biased. The transistor is in the active region and the collector current is depend on the emitter current. The transistor, which operates in this region is used for amplification.

2. FF – In this condition, both the junction is in forward biased. The transistor is in saturation and the collector current becomes independent of the base current. The transistors act like a closed switch.

3. RR – Both the current are in reverse biased. The emitter does not supply the majority charge carrier to the base and carriers current are not collected by the collector. Thus the transistors act like a closed switch.

4. RF – The emitter-base junction is in reverse bias and the collector-base junction is kept in forward biased. As the collector is lightly doped as compared to the emitter junction it does not supply the majority charge carrier to the base. Hence poor transistor action is achieved.

Bipolar Junction Transistor

The Bipolar Junction Transistor is a semiconductor device which can be used for switching or amplification bipolar transistors have the ability to operate within three different regions:

- Active Region – the transistor operates as an amplifier and Ic = β*Ib.
- Saturation – the transistor is "Fully-ON" operating as a switch and Ic = I(saturation).
- Cut-off – the transistor is "Fully-OFF" operating as a switch and Ic = 0.

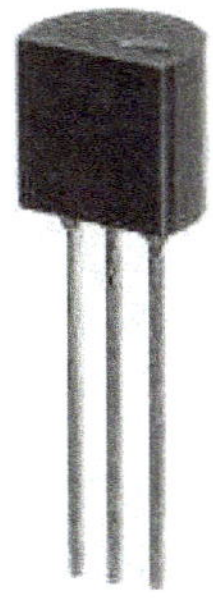

A Typical
Bipolar Transistor

The Bipolar Transistor basic construction consists of two PN-junctions producing three connecting terminals with each terminal being given a name to identify it from the other two. These three terminals are known and labelled as the Emitter (E), the Base (B) and the Collector (C) respectively.

Bipolar Transistors are current regulating devices that control the amount of current flowing through them from the Emitter to the Collector terminals in proportion to the amount of biasing voltage applied to their base terminal, thus acting like a current-controlled switch. As a small current flowing into the base terminal controls a much larger collector current forming the basis of transistor action.

The principle of operation of the two transistor types PNP and NPN, is exactly the same the only difference being in their biasing and the polarity of the power supply for each type.

Bipolar Transistor Construction

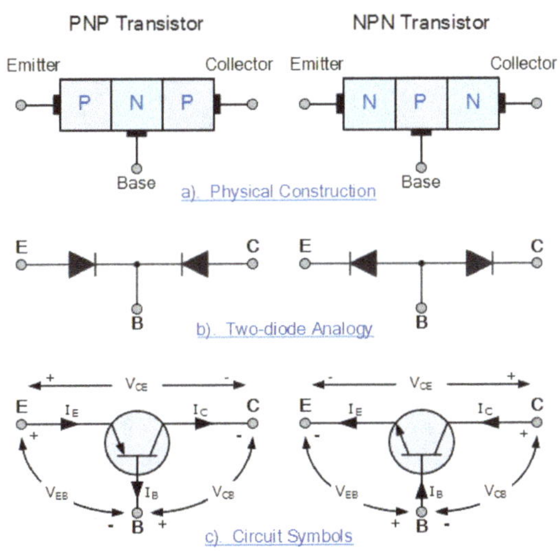

The construction and circuit symbols for both the PNP and NPN bipolar transistor are given above with the arrow in the circuit symbol always showing the direction of "conventional current flow" between the base terminal and its emitter terminal. The direction of the arrow always points from the positive P-type region to the negative N-type region for both transistor types, exactly the same as for the standard diode symbol.

Types of BJT

NPN Transistor

NPN transistor is one of the Bipolar Junction Transistor (BJT) types. The NPN transistor consists of two n-type semiconductor materials and they are separated by a thin layer of p-type semiconductor. Here the majority charge carriers are the electrons. The flowing of these electrons from emitter to collector forms the current flow in the transistor. Generally the NPN transistor is the most used type of bipolar transistors because the mobility of electrons is higher than the mobility of holes. The NPN transistor has three terminals – emitter, base and collector. The NPN transistor is mostly used for amplifying and switching the signals.

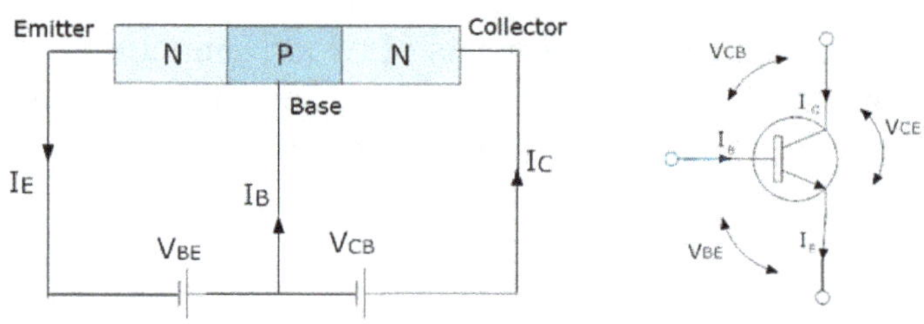

The above figure shows the symbol and structure of NPN transistor. In this structure we can observe the three terminals of transistor, circuit currents and voltage value representations.

NPN Transistor Circuit

The above figure shows the NPN transistor circuit with supply voltages and resistive loads. Here the collector terminal always connected to the positive voltage, the emitter terminal connected to the negative supply and the base terminal controls the ON/OFF states of transistor depending on the voltage applied to it.

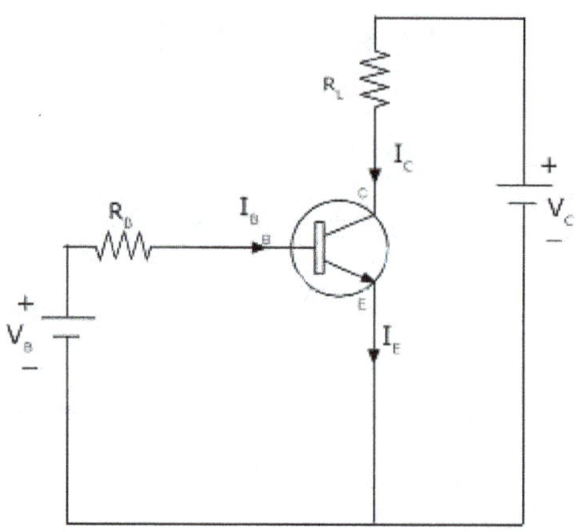

NPN Transistor Working

The working of NPN transistor is quite complex. In the above circuit connections we observed that the supply voltage VB is applied to the base terminal through the load RB. The collector terminal connected to the voltage VCC through the load RL. Here both the loads RB and RL can limit the current flow through the corresponding terminals. Here the base terminal and collector terminals always contain positive voltages with respect to emitter terminal.

If the base voltage is equal to the emitter voltage then the transistor is in OFF state. If the base voltage increases over emitter voltage then the transistor becomes more switched until it is in fully ON state. If the sufficient positive voltage is applied to the base terminal i.e. fully-ON state, then electrons flow generated and the current (IC) flows from emitter to the collector. Here the base terminal acts as input and the collector-emitter region acts as output.

To allow current flow between emitter and collector properly, it is necessary that the collector voltage must be positive and also greater than the emitter voltage of transistor. Some amount of voltage drop presented between base and emitter, such as 0.7V. So the base voltage must be greater than the voltage drop 0.7V otherwise the transistor will not operate. The equation for base current of a bipolar NPN transistor is given by:

$$I_B = (V_B - V_{BE})/R_B$$

Where,

I_B = Base current

V_B = Base bias voltage

V_{BE} = Input Base-emitter voltage = 0.7V

R_B = Base resistance

The output collector current in common emitter NPN transistor can be calculated by applying Kirchhoff's Voltage Law (KVL).

The equation for collector supply voltage is given as:

$$V_{CC} = I_C R_L + V_{CE}$$

From the above equation the collector current for common emitter NPN transistor is given as:

$$I_C = (V_{CC} - V_{CE})/R_L$$

In a common emitter NPN transistor the relation between collector current and emitter current is given as:

$$I_C = \beta\, I_B$$

In active region the NPN transistor acts as a good amplifier. In common emitter NPN transistor total current flow through the transistor is defined as the ratio of collector current to the base current IC/IB. This ratio is also called as "DC current gain" and it doesn't have any units. This ratio is generally represented with β and the maximum value of β is about 200. In common base NPN transistor the total current gain is expressed with the ratio of collector current to emitter current IC/IE. This ratio is represented with α and this value is generally equal to unity.

α, β and γ Relationship in NPN Transistor

Now let us see the relationship between the two ratio parameters α and β.

α = DC current gain for common base circuit = Output current/Input current

In common base NPN transistor output current is collector current (IC) and input current is emitter current (IE):

$$\alpha = I_C/I_E$$

This current gain (α) value is very close to unity but less than the unity.

The emitter current is the sum of small base current and large collector current:

$$I_E = I_C + I_B$$

$$I_B = I_E - I_C$$

from equation $\alpha = I_C/I_E$, the collector:

$$I_C = \alpha I_E$$

$$I_B = I_E - \alpha I_E$$

$$I_B = I_E(1\text{-}\alpha)$$

β = DC current gain for common emitter circuit = Output current/Input current

Here output current is collector current and input current is base current:

$$\beta = I_C/I_B$$

$$\beta = I_C/I_E(1\text{-}\alpha)$$

$$\beta = \alpha/(1\text{-}\alpha)$$

From the above equations the relationship between α and β can be expressed as:

$$\alpha = \beta\ (1\text{-}\alpha) = \beta/(\beta\text{+}1)$$

$$\beta = \alpha\ (1\text{+}\beta) = \alpha/(1\text{-}\alpha)$$

The β value may vary from 20 to 1000 for low power transistors which operate with high frequencies. But in general this β value can have the values in between the range of 50-200.

Now we will see the relationship between α, β and γ factors.

In common collector NPN transistor the current gain is defined as the ratio emitter current IE to base current IB. This current gain is represented with γ:

$$\gamma = I_E/I_B$$

emitter current:

$$I_E = I_C + I_B$$

$$\gamma = (I_C + I_B)/I_B$$

$$\gamma = (I_C/I_B) + 1$$

$$\gamma = \beta + 1$$

Hence the relationships between α, β and γ are given as below:

$$\alpha = \beta \,/\, (\beta+1), \beta = \alpha \,/\, (1-\alpha), \gamma = \beta + 1$$

Common Emitter Configuration

The common emitter configuration circuit is one of the three BJT configurations. These common emitter configuration circuits are used as voltage amplifiers. Generally the BJT transistors have three terminals but in the circuit connections we need to take any one terminal as common. So we use one of the three terminals as common terminal for both input and output actions. In this configuration we use emitter terminal as common terminal, thus it is named as common-emitter configuration.

This configuration is used as a single stage common emitter amplifier circuit. In this configuration base acts as input terminal, collector acts as output terminal and emitter as common terminal. The operation of this circuit starts with biasing the base terminal such that forward biasing the base-emitter junction. The small current in the base controls the current flow in the transistor. This configuration always operates in the linear region to amplify the signals at output side.

This common emitter amplifier gives the inverted output and can have very high gain. This gain value is influenced by temperature and bias current. The common-emitter amplifier circuit is mostly used configuration than other BJT configurations because of its high input impedance and low output impedance and also this configuration amplifier provides high voltage gain and power gain.

The current gain for this configuration is always greater than unity usually the typical value is about 50. These configuration amplifiers are mostly used in the applications where low frequency amplifier and radio frequency circuits are required. The circuit diagram for the common-emitter amplifier configuration is shown below.

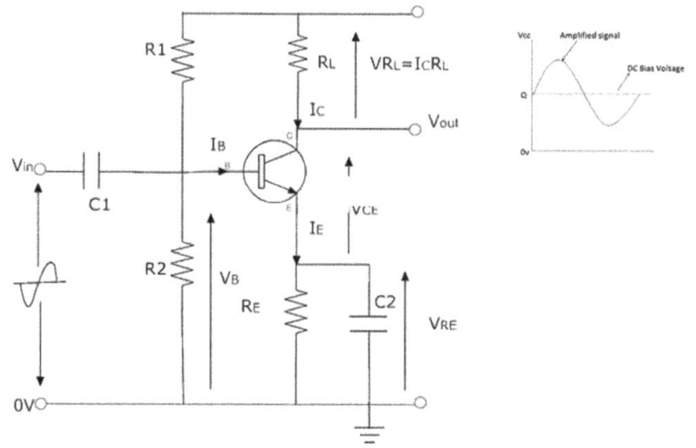

Output Characteristics of NPN Transistor

The family of output characteristics curves of a bipolar transistor is given below. The curves show the relationship between the collector current (IC) and the collector-emitter voltage (VCE) with the varying of base current (IB). the transistor is 'ON' only when at least a small amount of current and small amount of voltage is applied at its base terminal relative to emitter otherwise the transistor is in 'OFF' state.

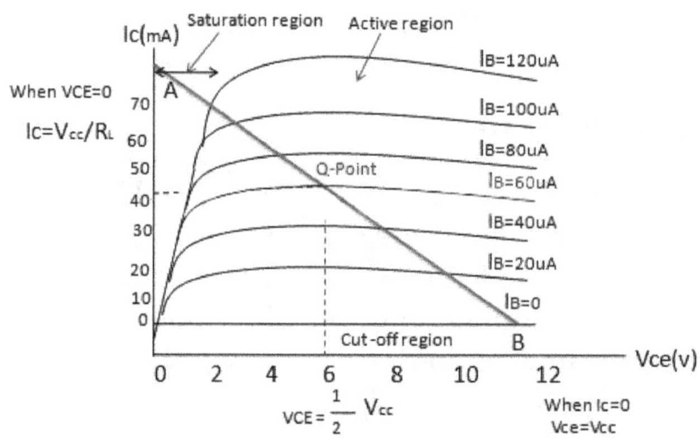

The collector current (IC) is mostly affected by the collector voltage (VCE) at 1.0V level but this IC value is not highly affected above this value. Already the emitter current is the sum of base and collector currents. i.e. IE =IC+ IB. The current flowing through the resistive load (RL) is equal to the collector current of the transistor. The equation for the collector current is given by,

$$I_C = (V_{CC} - V_{CE}) / R_L$$

The straight line indicates the 'Dynamic load line' which is connecting the points A (where VCE = 0) and B (where IC = 0). The region along this load line represents the 'active region' of the transistor.

The common emitter configuration characteristics curves are used to calculate the collector current when the collector voltage and base current is given. The load line (red line) is used to determine the Q-point in the graph. The slope of the load line is equal to the reciprocal of the load resistance. i.e. -1/RL.

NPN Transistor Applications

- NPN transistors are mainly used in switching applications.
- Used in amplifying circuit applications.
- Used in the Darlington pair circuits to amplify weak signals.
- NPN transistors are used in the applications where there is a need to sink a current.
- Used in some classic amplifier circuits, such as 'push-pull' amplifier circuits.
- In temperature sensors.
- Very High frequency applications.
- Used in logarithmic convertors.

PNP Transistor

PNP transistor is another type of Bipolar Junction Transistor (BJT). The structure of the PNP transistor is completely different from the NPN transistor. The two PN-junction diodes in the PNP transistor structure are reversed with respect to the NPN transistor, such as the two P-type doped semiconductor materials are separated by a thin layer of N-type doped semiconductor material.

In PNP transistor, the majority current carriers are holes and electrons are the minority current carriers. All the supply voltage polarities applied to the PNP transistor are reversed. In PNP, the current sinks in to the base terminal. The small base current in the PNP has the ability to control the large emitter-collector current because it is a current-controlled device.

The arrow for BJT transistors is always located on the emitter terminal and also it indicates the direction of conventional current flow. In PNP transistor this arrow indicates as 'pointing in' and the current direction in PNP is completely opposite to the NPN transistor. The structure of PNP transistor is completely opposite to the NPN transistor. But the characteristics and operation of the PNP transistor is almost same as NPN transistor with small differences. The symbol and structure for PNP transistor is shown below.

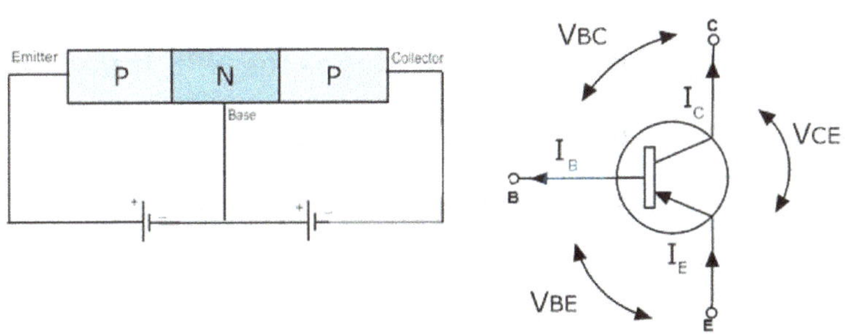

The above figure shows the structure and symbol of PNP Transistor. This transistor mainly consists of 3 terminals and they are Emitter (E), Collector (C) and Base (B). Here if you observe, the base current flows out of the base unlike NPN transistor. The emitter voltage is much positive with respect to base and collector.

PNP Transistor Working

The circuit connection of PNP transistor with supply voltages is given below. Here the base terminal has negative bias with respect to emitter and the emitter terminal has positive bias voltage with respect to both base and collector because of PNP transistor.

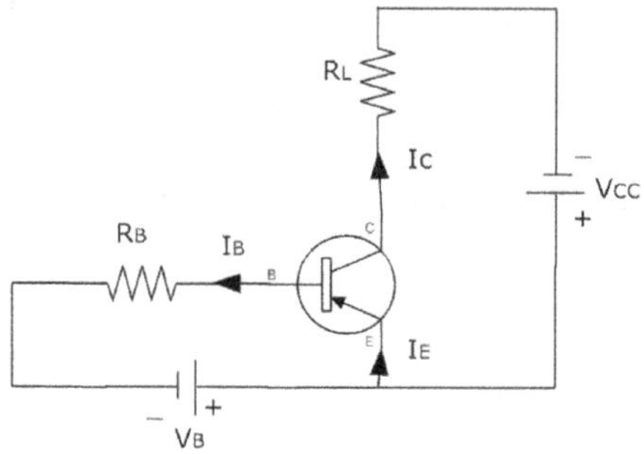

The polarities and current directions are reversed here compared to NPN transistor. If the transistor is connected to all the voltage sources as shown above then the base current flows through the transistor but here the base voltage needs to be more negative with respect to the emitter to operate transistor. Here the base- emitter junction acts as a diode. The small amount of current in the base controls the flowing of large current through emitter to collector region. The base voltage is generally 0.7V for Si and 0.3V for Germanium devices.

Here the base terminal acts as input and the emitter- collector region acts as output. The supply voltage V_{CC} is connected to the emitter terminal and a load resistor (R_L) is connected to the collector terminal. This load resistor (R_L) is used to limits the maximum current flow through the device. One more resistor (R_B) is connected to the base terminal which is used to limit the maximum current flow through the base terminal and also a negative voltage is applied to the base terminal. Here the collector current is always equal to the subtraction of base current from emitter current. Like NPN transistor, the PNP transistor also has the current gain value β. Now let us see the relation between the currents and current gain β.

The collector current (IC) is given by:

$$I_C = I_E - I_B$$

The DC current gain (β) for the PNP transistor is same as the NPN transistor.

DC current gain = β = Output current/Input current

Here output current is collector current and input current is base current:

$$\beta = I_C / I_B$$

From this equation we get;

$$I_B = I_C / \beta$$

$$I_C = \beta I_B$$

And also we define the current gain as,

Current gain = Collector current/ Emitter current (In common base transistor)

$$\alpha = I_C/I_E$$

The relation between α and β is given by:

$$\beta = \alpha / (1-\alpha) \text{ and } \alpha = \beta/(\beta+1)$$

The collector current in PNP transistor is given by:

$I_C = -\alpha\, I_E + I_{CBO}$ where I_{CBO} is the saturation current.

Since $I_E = -(I_C + I_B)$

$$I_C = -\alpha\, (-(I_C + I_B)) + I_{CBO}$$

$$I_C - \alpha\, I_C = \alpha\, I_B + I_{CBO}$$

$$I_C (1-\alpha) = \alpha\, I_B + I_{CBO}$$

$$I_C = (\alpha/(1-\alpha))\, I_B + I_{CBO}/(1-\alpha)$$

Since $\beta = \alpha / (1-\alpha)$

Now we get the equation for collector current:

$$I_C = \beta\, I_B + (1+\beta)\, I_{CBO}$$

The output characteristics of PNP transistor are same as NPN transistor characteristics. The small difference is that the PNP transistor characteristic curve rotates 1800 to calculate the reverse polarity voltages and current values. The dynamic load line also exists on the characteristic curve to calculate the Q-point value. The PNP transistors are also used in switching and amplifying circuits like NPN transistors.

PNP Transistor Example

Consider a PNP transistor, which is connected in the circuit with the supply voltages $V_B = 1.5V$, $V_E = 2V$, $+V_{CC} = 10V$ and $-V_{CC} = -10V$. And also this circuit connected with the resistors of RB = 200kΩ and R_B = 200kΩ and $R_E = R_C$ (or R_L) = 5kΩ. Now calculate the current gain values (α, β) of the PNP transistor.

Here,

$$V_B = 1.5V$$

$V_E = 2V$

$+V_{CC} = 10V$ and $-V_{CC} = -10V$

$R_B = 200k\Omega$

$R_E = R_C\,(\text{or } R_L) = 5k\Omega$

Base current,

$$I_B = V_B / R_B = 1.5/(200^*10^3) = 7.5uA.$$

Emitter current,

$$I_E = V_E / R_E = (10\text{-}2)/(5^*10^3) = 8/(5^*10^3) = 1.6mA$$

Collector current,

$$I_C = I_E - I_B = 1.6^*10^{-3} - 7.5^*10^{-6} = 1.59mA.$$

Now we have to calculate α and β values:

$$\alpha = I_C/I_E = 1.59*10^{-3}/1.6*10^{-3} = 0.995$$

$$\beta = I_C/I_B = 1.59*10^{-3}/7.5*10^{-6} = 212$$

Finally we get the current gain values of a considered PNP transistor are,

$$\alpha = 0.995 \text{ and } \beta = 212$$

BJT Transistor Matching

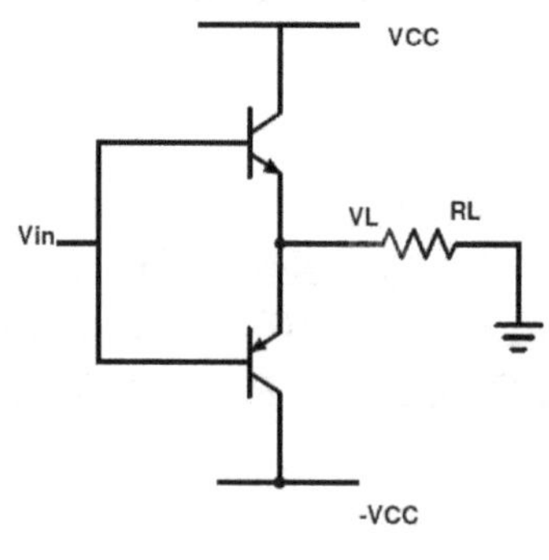

Transistor matching is nothing but connecting the both NPN and PNP transistors in a single design to generate high power. This structure is also called as "matched pair". The both NPN and

PNP transistors are called complementary transistors. Mainly these matched pair circuits are used in power amplifiers, such as class B amplifiers. If we connect the complementary transistors which are having the same characteristics then it is very useful to operate the output stages in motors and large machinery designs by producing high power continuously.

The NPN transistor conducts only in the positive half cycle of signal and the PNP transistor conducts only in the negative half cycle of the signal, due to this the device operates continuously. This continuous operation is very useful in the power motors to produce continuous power. The complementary transistors need to have same DC current gain (β) value. These matched pair circuits are used in motors controlling, robotics and power amplifier applications.

PNP Transistor Identification

Generally we identify the PNP transistors with their structure. We have some differences in the structures of both NPN and PNP transistors when compared. One more thing to identify the PNP transistor is generally the PNP transistor is in OFF for positive voltage and it is in ON when small output current and negative voltage at its base with respect to emitter. But to identify them with most efficiently we use some other technique by calculating the resistance between the three terminals, such as base, emitter and collector.

We have some standard resistance values for identifying the both NPN and PNP transistors. It is necessary to test each pair of terminals in both directions for resistance values so totally six tests are needed. This process is very much useful to identify the PNP transistor easily. Now we see the operation behavior of each pair of terminals.

- Emitter-Base Terminals: The emitter-base region acts as a diode but it conducts only in one direction.
- Collector-Base Terminals: The collector-base region also acts as diode which conducts I only one direction.
- Emitter-Collector Terminals: The emitter- collector region looks like a diode but it will not conduct in either direction.

Now let us see the resistance value table to identify both NPN and PNP transistors as shown in the following table.

Between Transistor Terminals	PNP	NPN	
Collector	Emitter	R_{HIGH}	R_{LOW}
Collector	Base	R_{LOW}	R_{LOW}
Emitter	Collector	R_{HIGH}	R_{LOW}
Emitter	Base	R_{LOW}	R_{LOW}
Base	Collector	R_{HIGH}	R_{LOW}
Base	Emitter	R_{HIGH}	R_{LOW}

PNP Transistor as a Switch

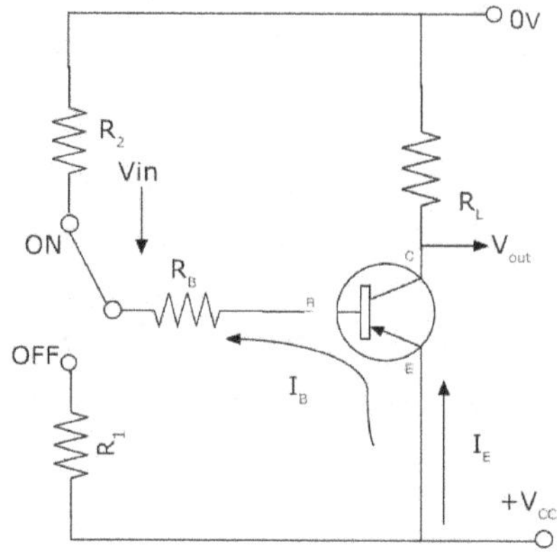

The circuit in the above figure shows the PNP transistor as a switch. The operation of this circuit is very simple, if the input pin of transistor (base) is connected to ground (i.e. negative voltage) then the PNP transistor is in 'ON', now the supply voltage at emitter conducts and the output pin pulled up to the larger voltage. If the input pin connected to the high voltage (i.e. positive voltage) then the transistor is 'OFF', so the output voltage has to be low (zero). This operation shows the switching conditions of a PNP transistor due to their ON and OFF states.

Applications

- PNP transistors are used to source current, i.e. current flows out of the collector.
- PNP transistors are used as switches.
- These are used in the amplifying circuits.
- PNP transistors are used when we need to turnoff something by push a button. i.e. emergency shutdown.
- Used in Darlington pair circuits.
- Used in matched pair circuits to produce continuous power.
- Used in heavy motors to control current flow.
- Used in robotic applications.

Field-effect Transistor

The field effect transistor, FET is a key active device for the electronics industry.

The FET used in many circuits constructed from discrete components in areas from RF technology to power control and switching to general amplification.

However the major use for the field effect transistor, FET is within integrated circuits. In this application FET circuits are able to consume only very small levels of power, and this enables the huge very large scale integrated circuits to operate. If bipolar technology was used the power consumption would be orders of magnitude greater and the power generated far too large to accommodate within a single integrated circuit.

The concept of the field effect transistor is based around the concept that charge on a nearby object can attract charges within a semiconductor channel. It essentially operates using an electric field effect - hence the name.

The FET consists of a semiconductor channel with electrodes at either end referred to as the drain and the source.

A control electrode called the gate is placed in very close proximity to the channel so that its electric charge is able to affect the channel.

In this way, the gate of the FET controls the flow of carriers (electrons or holes) flowing from the source to drain. It does this by controlling the size and shape of the conductive channel.

The semiconductor channel where the current flow occurs may be either P-type or N-type. This gives rise to two types or categories of FET known as P-Channel and N-Channel FETs.

In addition to this, there are two further categories. Increasing the voltage on the gate can either deplete or enhance the number of charge carriers available in the channel. As a result there are enhancement mode FET and depletion mode FETs.

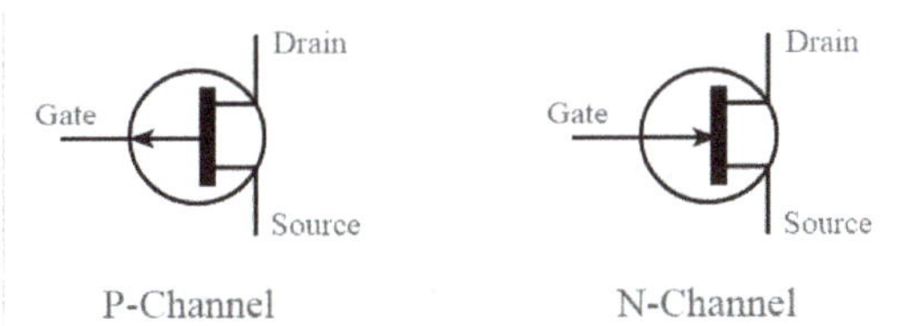

Junction FET circuit symbol

As it is only the electric field that controls the current flowing in the channel, the device is said to be voltage operated and it has a high input impedance, usually many megohms. This can be a distinct advantage over the bipolar transistor that is current operated and has a much lower input impedance.

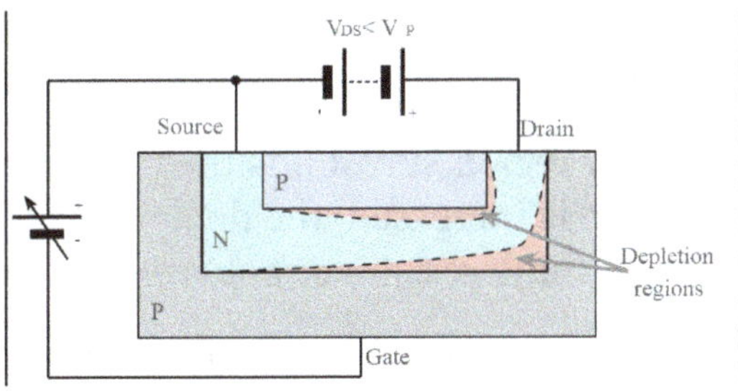

Junction FET working below saturation

FET Circuits

Field effect transistors are widely used in all forms of circuit from those used in circuits with discrete components, to those employed in integrated circuits.

Note on field effect transistor circuit design:

The field transistor transistors can be used in many types of circuits although the three basic configurations are common source, common drain (source follower) and common gate. The circuit design itself if fairly straightforward and can be undertaken quite easily.

Field Effect Transistor Types

There are many ways to define the different types of FET that are available. They may be categorised in a number of ways, but some of the major types of FET can be covered in the tree diagram below.

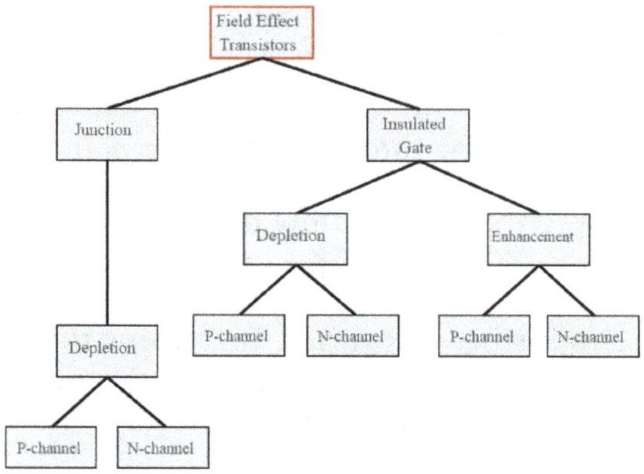

Types of FET

Junction Field-Effect Transistor (JFET)

In its simplest form the junction field-effect transistor starts with nothing more than a bar of doped silicon that behaves as a resistor. By convention, the terminal into which current is injected is called the source terminal, since, as far as the FET is concerned, current originates from this terminal. The other terminal is called the drain terminal Current flow between source and drain is related to the drain-source voltage by the resistance of the intervening material. In figure, p-type regions have been diffused into the n-type substrate of figure leaving an n-type channel between the source and drain. (A complementary p-type device is made by reversing all of the material types.) These p-type regions will be used to control the current flow between the source and the drain and are thus called gate regions.

As with any p-n junction, a depletion region surrounds the p-n junctions when the junctions are reverse biased. As the reverse voltage is increased, the depletion regions spread into the channel until they meet, creating an almost infinite resistance between the source and the drain.

If an external voltage is applied between source and drain Increases, the depletion regions again spread into the channel because of the voltage drop in the channel which reverse biases the junctions. As VDS is increased, the depletion regions grow until they meet, whereby any further Increase in voltage is counterbalanced by an increase in the depletion region toward the drain. There is an effective increase in channel resistance that prevents any further increase in drain current. The drain-source voltage that causes this current limiting condition is called the "pinch off" Voltage (V_p). A further increase in drain-source voltage produces only a slight increase in drain current.

The variation in drain current (I_D) with drain-source voltage (V_{DS}) at zero gate-source voltage (V_{GS}) is shown in Figure. In the low-current region, the drain current is linearly related to V_{DS}. As ID increases, the "channel" begins to deplete and the slope of the ID curve decreases. When the V_{DS} is equal to V_p, ID "saturates" and stays relatively constant until drain-to-gate avalanche, $V_{BR\ (DSS)}$ is reached. If a reverse voltage is applied to the gates, channel pinch-off occurs at a lower ID level because the depletion region spread caused by the reverse-biased gates adds to that produced by V_{DS}. Thus reducing the maximum current for any value of V_{DS}.

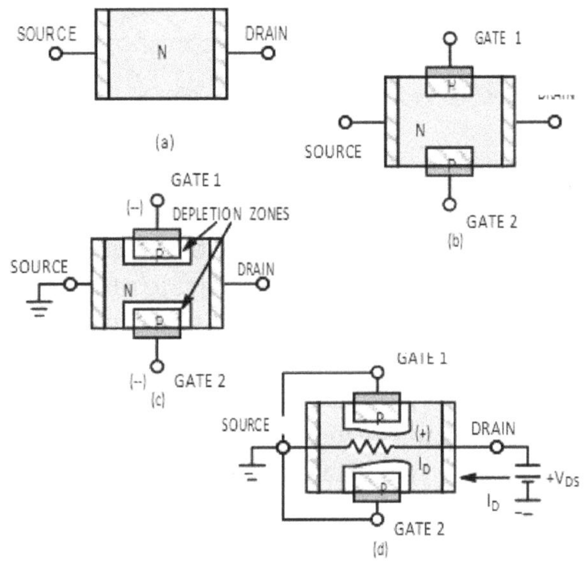

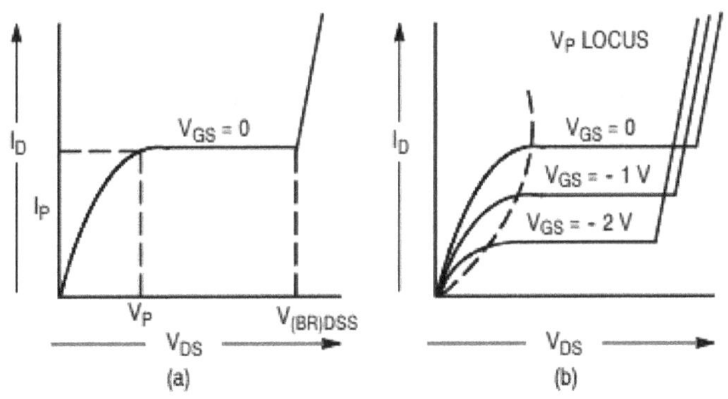

Drain Current Characteristics

Due to the difficulty of diffusing impurities into both sides of a semiconductor wafer, a single ended geometry is normally used instead of the two-sided structure discussed above. Diffusion for this geometry is from one side only. The substrate is of p-type material onto which an n-type channel is grown epitaxially. A p-type gate is then diffused into the n-type epitaxial channel. Contact metallization completes the structure.

The substrate, which functions as Gate 2 of Figure, is of relatively low resistivity material to maximize gain. For the same purpose, Gate 1 is of very low resistivity material, allowing the depletion region to spread mostly into the n-type channel. In most cases the gates are internally connected together. A tetrode device can be realized by not making this internal connection.

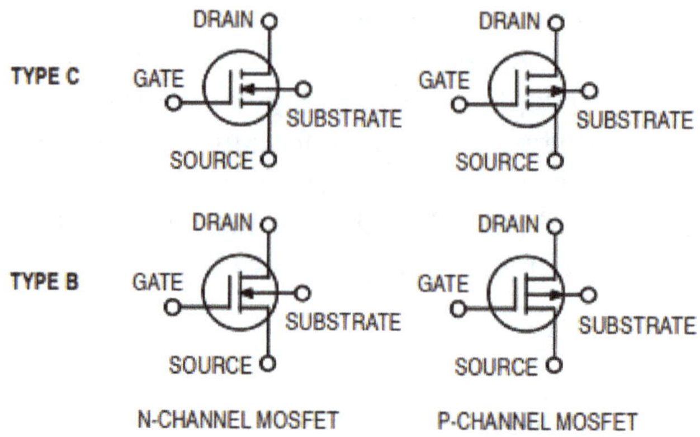

Mos Field-Effect Transistors (MOSFET)

The metal-oxide-semiconductor (MOSFET) operates with a slightly different control mechanism than the JFET. Figure shows the development. The substrate may be high resistivity p-type material, as for the 2N4351. This time two separate low-resistivity n-type regions (source and drain) are diffused into the substrate as shown in figure. Next, the surface of the structure is covered with an insulating oxide layer and a nitride layer. The oxide layer serves as a protective coating for the FET surface and to insulate the channel from the gate. However the oxide is subject to contamination by sodium ions which are found in varying quantities in all environments. Such contamination results in long term instability and changes in device characteristics.

Silicon nitride is impervious to sodium ions and thus is used to shield the oxide layer from contamination. Holes are cut into the oxide and nitride layers allowing metallic contact to the source and drain. Then, the gate metal area is overlaid on the insulation, covering the entire channel region and, simultaneously, metal contacts to the drain and source are made as shown in figure. The contact to the metal area covering the channel is the gate terminal. Note that there is no physical penetration of the metal through the oxide and nitride into the substrate. Since the drain and source are isolated by the substrate, any drain-to-source current in the absence of gate voltage is extremely low because the structure is analogous to two diodes connected back to back. The metal area of the gate forms a capacitor with the insulating layers and the semiconductor channel. The metal area is the top plate; the substrate material and channel are the bottom plate.

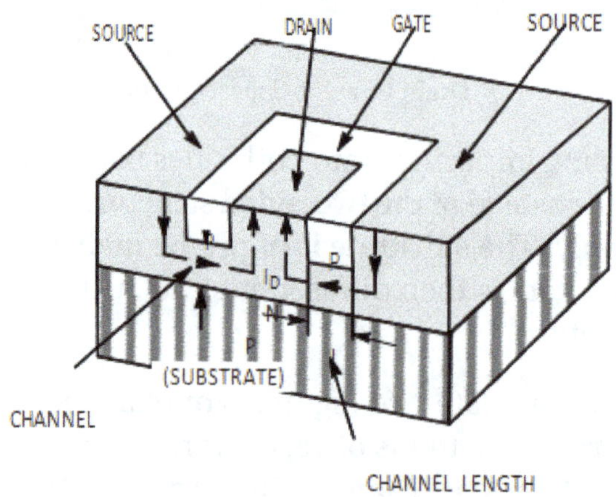

Junction FET with Single-Ended Geometry

For the structure of figure, consider a positive gate potential.Positive charges at the metal side of the metal-oxide capacitor induce a corresponding negative charge at the semiconductor side. As the positive charge at the gate is increased, the negative charge "induced" in the semiconductor increases until the region beneath the oxide effectively becomes an n-type semiconductor region, and current can flow between drain and source through the "induced" channel. In other words, drain current flow is "enhanced" by the gate potential. Thus drain current flow can be modulated by the gate voltage; i.e. the channel resistance is directly related to the gate voltage. The n-channel structure may be changed to a p-channel device by reversing the material types.

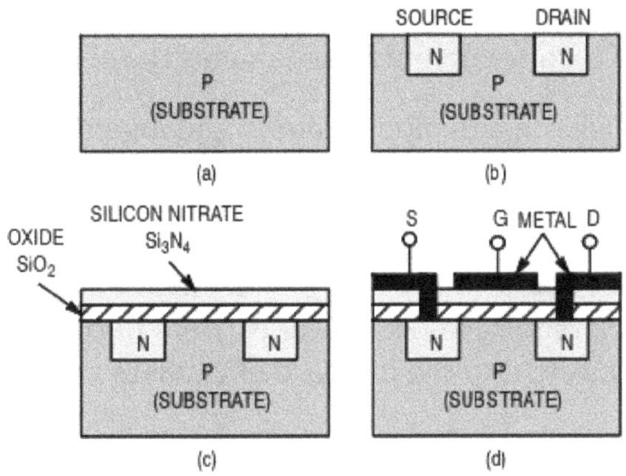

Development of Enhancement-Mode N-Channel MOSFET

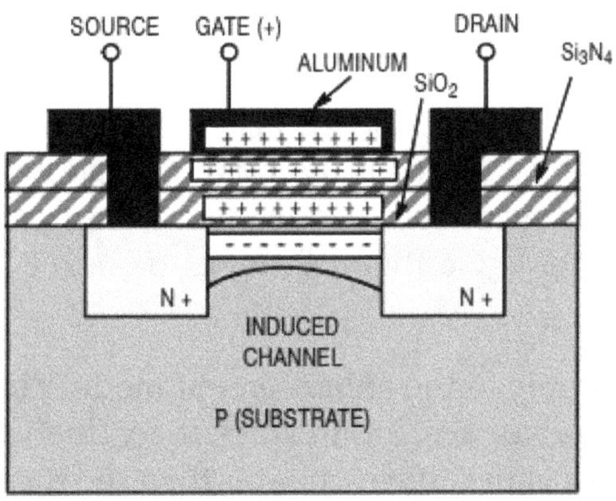

Channel Enhancement. Application of Positive Gate Voltage Causes Redistribution of Minority Carriers in the Substrate and Results in the Formation of a Conductive Channel Between Source and Drain

An equivalent circuit for the MOSFET is shown in. Here, $C_{g(ch)}$ is the distributed gate-to-channel capacitance representing the nitride-oxide capacitance. C_{gs} is the gate-source capacitance of the metal gate area overlapping the source, while C_{gd} is the gate-drain capacitance of the metal gate area overlapping the drain. $C_{d(sub)}$ and $C_{s(sub)}$ are junction capacitances from drain to substrate and source to substrate. Y_{fs} is the transadmittance between drain current and gate-source voltage. The modulated channel resistance is rds. R_D and R_S are the bulk resistances of the drain and source.

The input resistance of the MOSFET is exceptionally high because the gate behaves as a capacitor with very low leakage ($r_{in} \approx 10_{14}\ \Omega$). The output impedance is a function of r_{ds} (which is related to the gate voltage) and the drain and source bulk resistances (R_D and R_S).

To turn the MOSFET "on", the gate-channel capacitance, $C_{g(ch)}$, and the Miller capacitance, C_{gd}, must be charged. In turning "on", the drain-substrate capacitance, $C_{d(sub)}$, must be discharged. The resistance of the substrate determines the peak discharge current for this capacitance.

The FET just described is called an enhancement-type MOSFET. A depletion-type MOSFET can be made in the following manner: Starting with the basic structure of a moderate resistivity n-channel is diffused between the source and drain so that drain current can flow when the gate potential is at zero volts. In this manner, the MOSFET can be made to exhibit depletion characteristics. For positive gate voltages, the structure enhances in the same manner as the device of. With negative gate voltage, the enhancement process is reversed and the channel begins to deplete of carriers as seen in. As with the JFET, drain-current flow depletes the channel area nearest the drain first.

The structure of, therefore, is both a depletion-mode and an enhancement-mode device.

Modes of Operation

There are two basic modes of operation of FET's — depletion and enhancement. Depletion mode, as previously mentioned, refers to the decrease of carriers in the channel due to variation in gate voltage. Enhancement mode refers to the increase of carriers in the channel due to application of gate voltage.

The basic differences between these modes can most easily be understood by examining the transfer characteristics of Figure. The depletion-mode device has considerable drain-current flow for zero gate voltage. Drain current is reduced by applying a reverse voltage to the gate terminal. The depletion-type FET is not characterized with forward gate voltage.

The depletion/enhancement mode type device also has considerable drain current with zero gate voltage. This type device is defined in the forward region and may have usable forward characteristics for quite large gate voltages. Notice that for the junction FET, drain current may be enhanced by forward gate voltage only until the gate-source p-n junction becomes forward biased.

The third type of FET operates only in the enhancement mode. This FET has extremely low drain current flow for zero gate-source voltage. Drain current conduction occurs for a V_{GS} greater than some threshold value, $V_{GS(th)}$. For gate voltages greater than the threshold, the transfer characteristics are similar to the depletion/enhancement mode FET.

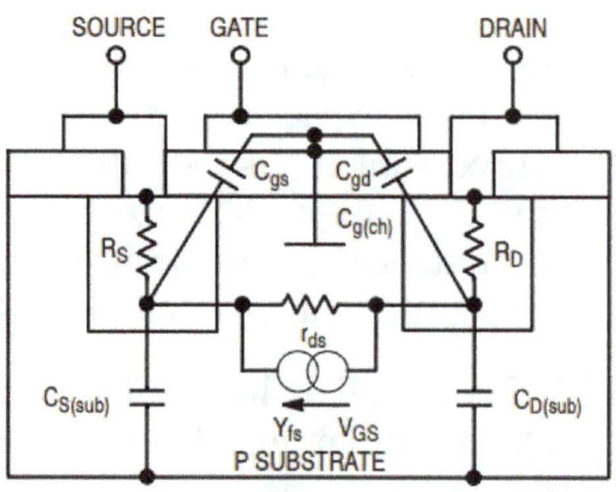

Equivalent Circuit of Enhancement Mode MOSFET

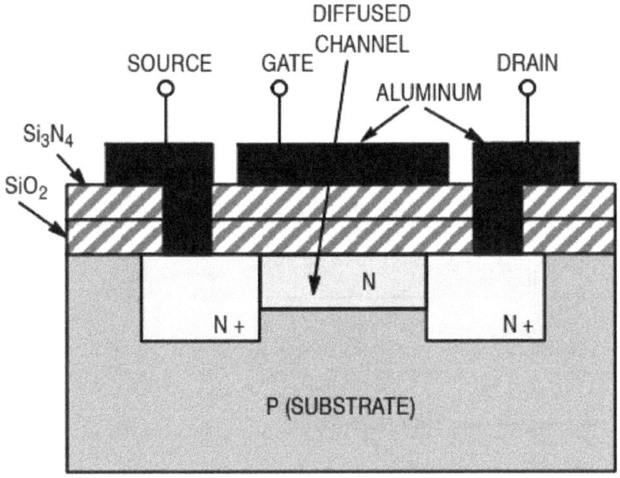

Depletion Mode MOSFET Structure. This Type of Device May Be Designed to Operate in Both the Enhancement and Depletion Modes

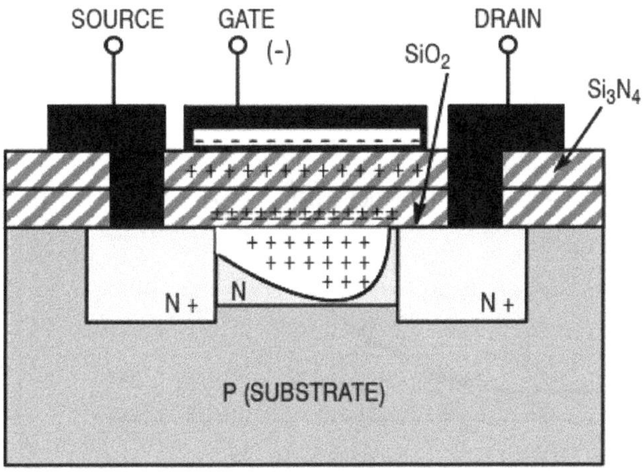

Channel Depletion Phenomenon. Application of Negative Gate Voltage Causes Redistribution of Minority Carriers in Diffused Channel and Reduces Effective Channel Thickness. This Results in Increased Channel Resistance

Electrical Characteristics

Because the basic mode of operation for field-effect devices differs greatly from that of conventional junction transistors, the terminology and specifications are necessarily different. An understanding of FET terminology and characteristics are necessary to evaluate their comparative merits from data-sheet specifications.

Static Characteristics

Static characteristics define the operation of an active device under the influence of applied dc operating conditions. Of primary interest are those specifications that indicate the effect of a control signal on the output current. The V_{GS} -- ID transfer characteristics curves are illustrated in for the three types of FETs. lists the data-sheet specifications normally employed to describe these curves, as well as the test circuits that yield the indicated specifications.

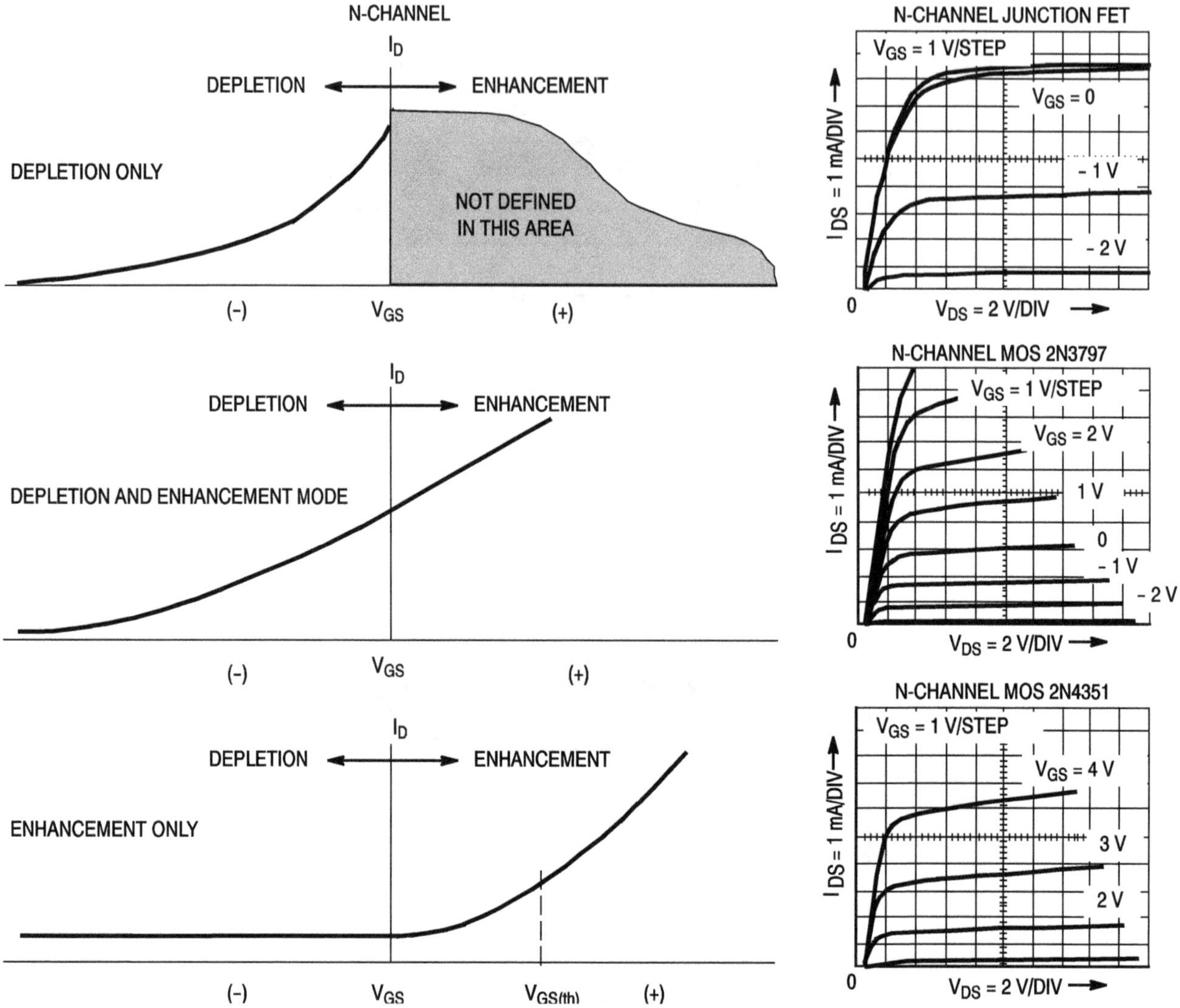

Transfer Characteristics and Associated Scope Traces for the Three FET Types

Of additional interest is the special case of tetrode connected devices in which the two gates are separately accessible for the application of a control signal. The pertinent specifications for a junction tetrode are those which define drain-current cut off when one of the gates is connected to the source and the bias voltage is applied to the second gate. These are usually specified as $V_{G1S(off)}$, Gate 1 — source cut off voltage (with Gate 2 connected to source), and $V_{G2S(off)}$, Gate 2 — source cut off voltage (with Gate 1 connected to source). The gate voltage required for drain current cut off with one of the gates connected to the source is always higher than that for the triode-connected case where both gates are tied together.

Reach-through voltage is another specification uniquely applicable to tetrode-connected devices. This defines the amount of difference voltage that may be applied to the two gates before the depletion region of one spreads into the junction of the other — causing an increase in gate current to some small specified value. Obviously, reach-through is an undesirable condition since it causes a decrease in input resistance as a result of an increased gate current, and large amounts of reach-through current can destroy the FET.

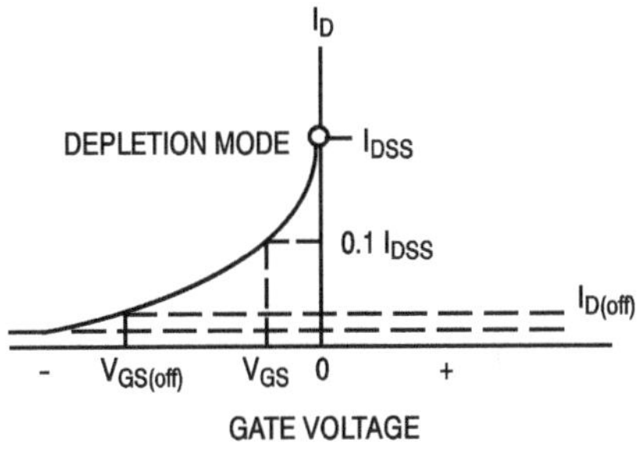

DEPLETION MODE JFETs

CHARACTERISTIC	DESCRIPTION
I_{DSS} @ $V_{GS} = 0$, $V_P < V_{DS} < V_{(BR)DSS}$	Zero-gate-voltage drain current. Represents maximum drain current.
$V_{GS(off)}$ @ $I_D = 0.001\ I_{DSS}$, $V_P < V_{DS} < V_{(BR)DSS}$	Gate voltage necessary to reduce I_D to some specified negligible value at the recommended V_{DS} i.e. cutoff.
V_{GS} @ $I_D = 0.1\ I_{DSS}$, $V_P < V_{DS} < V_{(BR)DSS}$	Gate voltage for a specified value of I_D between I_{DSS} and I_{DS} at cutoff - normally 0.1 I_{DSS}.

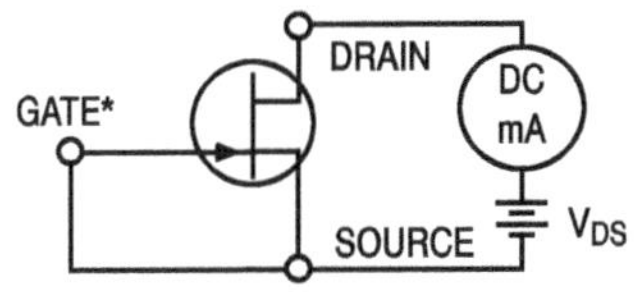

TEST CIRCUIT FOR I_{DSS}

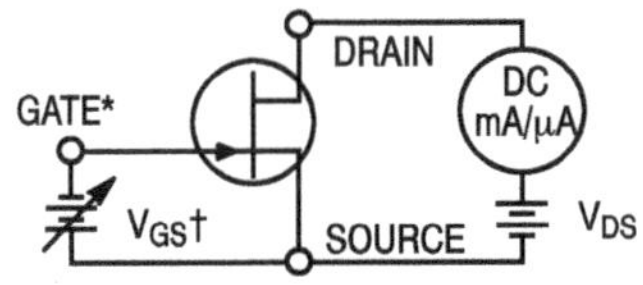

TEST CIRCUIT FOR V_{GS} AND $V_{GS(off)}$
* GATES INTERNALLY CONNECTED
† ADJUST FOR DESIRED I_D

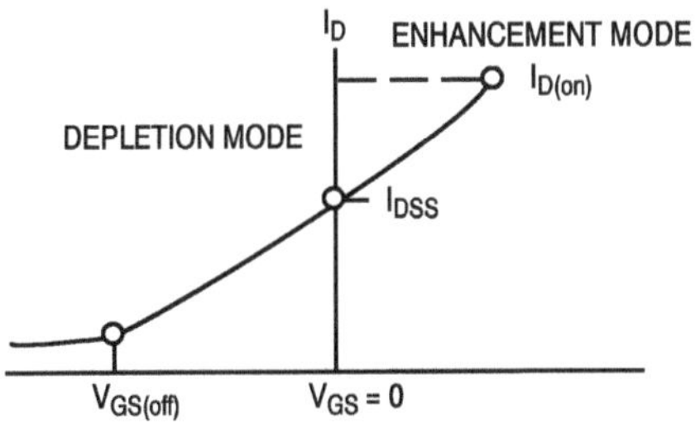

DEPLETION/ENHANCEMENT MODE MOSFETs

CHARACTERISTIC	DESCRIPTION
$I_{D(on)}$ @ $V_{GS} > 0$, $V_P < V_{DS} < V_{(BR)DSS}$	An arbitrary current value (usually near max rated current) that locates a point in the enhancement operation mode.
I_{DSS} @ $V_{GS} = 0$, $V_P < V_{DS} < V_{(BR)DSS}$	Zero-gate-voltage drain current.
$V_{GS(off)}$ @ $I_D = 0.001\ I_{DSS}$	Voltage necessary to reduce I_D to some specified negligible value at the recommended V_{DS}, i.e. cutoff.

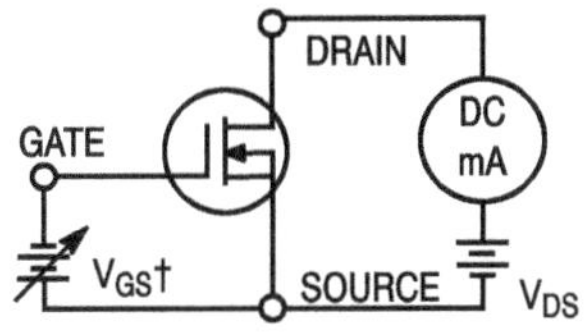

TEST CIRCUIT FOR $I_{D(on)}$
† ADJUST FOR DESIRED I_D, NORMALLY NEAR MAX-RATED I_D

TEST CIRCUIT FOR I_{DSS} AND $V_{GS(off)}$

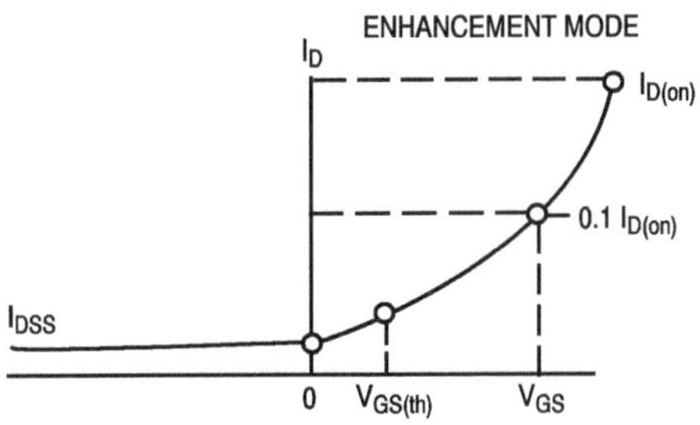

ENHANCEMENT MODE MOSFETs

CHARACTERISTIC	DESCRIPTION
$I_{D(on)}$ @ $V_{GS} > 0$, $V_P < V_{DS} < V_{(BR)DSS}$	An arbitrary current value (usually near max rated current) that locates a point in the enhancement operation mode.
V_{GS} @ 0.1 $I_{D(on)}$	Gate-source voltage for a specified drain current of 0.1 $I_{D(on)}$.
$V_{GS(th)}$ @ $I_D = 0.001$ $I_{D(on)}$ or less	Gate cutoff or turn-on voltage.
I_{DSS} @ $V_{GS} = 0$, $V_P < V_{DS} < V_{(BR)DSS}$	Leakage drain current.

$I_{D(on)}$ TEST CIRCUIT
SAME AS FOR DEPLETION/ ENHANCEMENT

V_{GS} TEST CIRCUIT
SAME AS FOR $I_{D(on)}$

$V_{GS(th)}$ TEST CIRCUIT
SAME AS FOR $V_{GS(off)}$ FOR JFET EXCEPT REVERSE V_{GS} BATTERY POLARITY

I_{DSS} TEST CIRCUIT
SAME AS FOR JFET

Static Characteristics for the Three FET Types Are Defined by the Above Curves, Tables, and Test Circuits

Gate Leakage Current

Of interest to circuit designers is the input resistance of an active component. For FETs, this characteristic is specified in the form of IGSS — the reverse-bias gate-to-source current with the drain shorted to the source. As might be expected, because the leakage current across a reverse-biased p-n junction (in the case of a JFET) and across a capacitor (in the case of a MOSFET) is very small, the input resistance is extremely high. At a temperature of 25 °C, the JFET input resistance is hundreds of megohms while that of a MOSFET is even greater. For junction devices, however, input resistance may decrease by several orders of magnitude as temperature is raised to 150 °C. Such devices, therefore, have gate-leakage current specified at two temperatures. Insulated-gate FETs are not drastically affected by temperature, and their input resistance remains extremely high even at elevated temperatures.

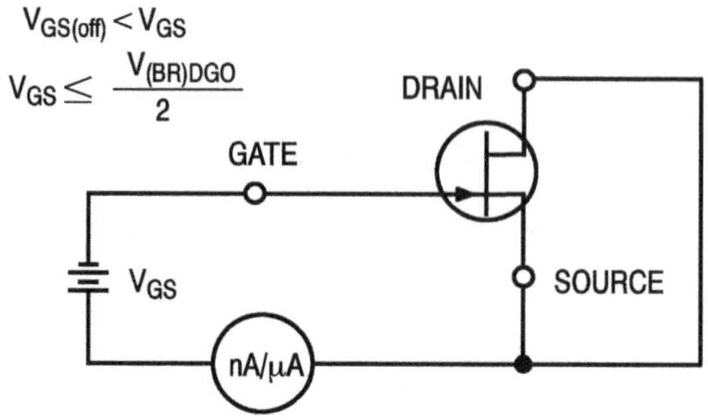

Test Circuit for Leakage Current

Gate leakage current may also be specified as IGDO (leakage between gate and drain with the source open), or as IGSO (leakage between gate and source with the drain open). These usually result in lower values of leakage current and do not represent worst-case conditions. The IGSS specification, therefore, is usually preferred by the user.

Voltage Breakdown

A variety of specifications can be used to indicate the maximum voltage that may be applied to various elements of a FET. Among those in common use are the following:

$V_{(BR)GSS}$ = Gate-to-source breakdown voltage

$V_{(BR)DGO}$ = Drain-to-gate breakdown voltage

$V_{(BR)DSX}$ = Drain-to-source breakdown voltage

(normally used only for MOSFETs)

In addition, there may be ratings and specifications indicating the maximum voltages that may be applied between the individual gates and the drain and source (for tetrode connected devices). Obviously, not all of these specifications are found on every data sheet since some of them provide the same information in somewhat different form.

In junction FETs, the maximum voltage that may be applied between any two terminals is the lowest voltage that will lead to breakdown or avalanche of the gate junction. To measure $V_{(BR)GSS}$, an increasingly higher reverse voltage is applied between the gate and the source. Junction breakdown is indicated by an increase in gate current (beyond IGSS) which signals the beginning of avalanche.

Some reflection will reveal that for junction FETs, the $V_{(BR)DGO}$ specification really provides the same information as $V_{(BR)GSS}$. For this measurement, an increasing voltage is applied between drain and gate. When this applied voltage becomes high enough, the drain-gate junction will go into avalanche, indicated either by a significant increase in drain Current or by an increase in gate current (beyond I_{DGO}). For both $V_{(BR)DGO}$ and $V_{(BR)GSS}$ specifications, breakdown should normally occur at the same voltage value.

From Figure it is seen that avalanche occurs at a lower value of V_{DS} when the gate is reverse biased than for the zero-bias condition. This is caused by the fact that the reverse-bias gate voltage adds to the drain voltage, thereby Increasing the effective voltage across the junction. The maximum amount of drain-source voltage that may be applied $V_{DS(max)}$ is, therefore, equal to $V_{(BR)DGO}$ minus VGS, which indicates avalanche with reverse bias gate voltage applied.

For MOSFETs, the breakdown mechanism is somewhat different. Consider, for example, the enhancement-mode structure of Figure. Here, the gate is completely insulated from the drain, source, and channel by an oxide-nitride layer. The breakdown voltage between the gate and any of the other elements, therefore, is dependent on the thickness and purity of this insulating layer, and represents the voltage that will physically puncture the layer. Consequently, the voltage must be specified separately.

The drain-to-source breakdown is a different matter. For enhancement mode devices, with the gate connected to the source (the cut off condition) and the substrate floating, there is no effective channel between drain and source and the applied drain-source voltage appears across two opposed series diodes, represented by the source-to-substrate and substrate-to-drain junctions. Drain current remains at a very low level (picoamperes) as drain voltage is increased until the drain voltage reaches a value that causes reverse (avalanche) breakdown of the diodes. This particular condition, represented by V(BR)DSS, is indicated by an Increase in ID above the IDSS level, as shown in figure.

For depletion/enhancement mode devices, the $V_{(BR)DSS}$ symbol is sometimes replaced by $V_{(BR)DSX}$. Note that the principal difference between the two symbols is the replacement of the last subscript s with the subscript x. Whereas the s normally indicates that the gate is shorted to the source, the x indicates that the gate is biased to cut off or beyond. To achieve cut off in these devices, a depleting bias voltage must be applied to the gate.

An important static characteristic for switching FETs is the "on" drain-source voltage VDS(on). This characteristic for the MOSFETs is a function of VGS, and resembles the $V_{CE(sat)}$ versus I_B characteristics of junction transistors. The curve for these characteristics can be used as a design guide to determine the minimum gate voltage necessary to achieve a specified output logic level.

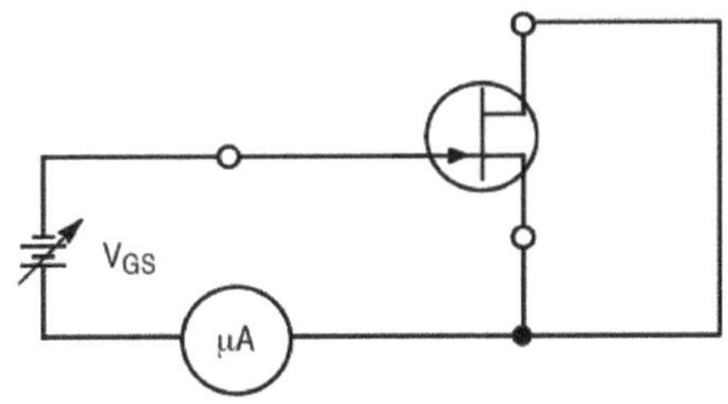

$V_{(BR)GSS}$ Test Circuit

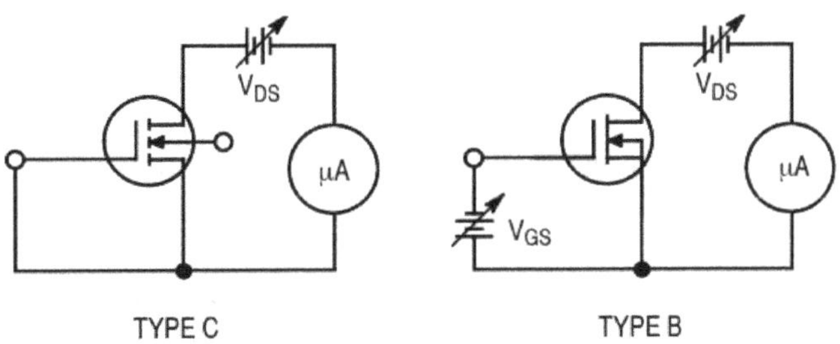

$V_{(BR)DSS}$ and $V_{(BR)DSX}$ Test Circuit

Dynamic Characteristics

Unlike the static characteristics, the dynamic characteristics of field-effect transistors apply equally to all FETs. The conditions and presentation of the dynamic characteristics, however, depend largely upon the intended application. For example, the following table indicates the dynamic characteristics needed to adequately describe a FET for various applications.

Audio	RF-IF	Switching	Chopper
yfs (1kHz)	yfs (1 kHz)		
C_{iss}	C_{iss}	C_{iss}	C_{iss}
C_{rss}	C_{rss}	C_{rss}	C_{rss}
y_{os} (1kHz)	GP	$Cd_{(sub)}$	$Cd_{(sub)}$
NF	Re(y_{is}) (HF)	$r_{ds(on)}$	$r_{ds(on)}$
	Re($_{yos}$) (HF)	t_{d1}, t_{d2}	
	NF	t_r, t_f	

Y_{fs} The forward transadmittance is a key dynamic characteristic for field-effect transistors. It serves as a basic design parameter in audio and rf circuits and is a widely accepted figure of merit for devices.

Because field-effect transistors have many characteristics similar to those of vacuum tubes, and because many engineers still are more comfortable with tube parameters, the symbol gm used for tube transconductance is often Specified instead of y_{fs}. In addition to gm, some data sheets show gfs while others even show g_{21}.

Regardless of the symbol used, y_{fs} defines the relation between an input signal voltage and an output signal current:

$$y_{fs} = \Delta I_D / \Delta V_{GS} \Big|_{V_{DS} = K}$$

The unit is the mho — current divided by voltage. Figure is a typical y_{fs} test circuit for a junction FET.

As a characteristic of all field-effect devices, y_{fs} is specified at 1 kHz with a V_{DS} the same as that for which ID(on) or IDSS is characterized. Since y_{fs} has both real and imaginary components, but is dominated by the real component at low frequency, the 1 kHz characteristic is given as an absolute magnitude and indicated as $|y_{fs}|$.

It is interesting to note that y_{fs} varies considerably with I_D due to nonlinearity in the I_D -- V_{GS} characteristics. This variation, for a typical n-channel, JFET is illustrated in Figure. Obviously, the operating point must be carefully selected to provide the desired y_{fs} and signal swing.

For tetrode-connected FETs, three y_{fs} measurements are usually specified on data-sheet tables. One of these, with the two gates tied together, provides a y_{fs} value for the condition where a signal is applied to both gates simultaneously; the others provide the yfs for the two gates individually.

Generally, with the two gates tied together, y_{fs} is higher and more gain may be realized in a given circuit. Because of the increased capacitance, however, gain-bandwidth product is much lower.

For rf field-effect transistors, an additional value of y_{fs} is sometimes specified at or near the highest frequency of operation. This value should also be measured at the same Voltage conditions as those used for $I_{D(on)}$ or I_{DSS}. Because of the importance of the imaginary component at radio frequencies, the high frequency y_{fs} specification should be a complex representation, and should be given either in the specifications table or by means of curves showing typical variations, as in Figure for the MPF102 JFET.

The real portion of this high-frequency y_{fs}, Re(yfs) or G_{21}, is usually considered a significant figure of merit.

y_{os} Another FET parameter that offers a direct vacuum tube analogy is yos, the output admittance:

$$y_{OS} = \Delta I_D / \Delta V_{DS} \Big|_{V_{GS} = K}$$

In this case, the analogous tube parameter is r_p — i.e., $y_{os} = 1/r_p$. For depletion mode devices, y_{os} is measured with gate and source grounded. For enhancement mode units, it is measured at some specified VGS that permits substantial drain-current flow.

As with y_{fs}, many expressions are used for y_{os}. In addition to the obvious parallels such as y22, g_{os}, and g22, it is also sometimes specified as r_d, where rd = $1/y_{os}$

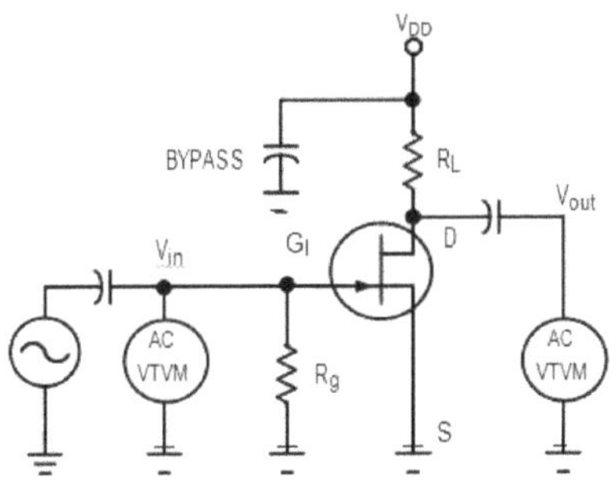

$$Y_{fsl} = \frac{V_{out}}{V_{in} R_L}$$

Rg typically 1 MΩ
RL of such value as to cause negligible dc drop at idss

Typical y_{fs} Test Circuit

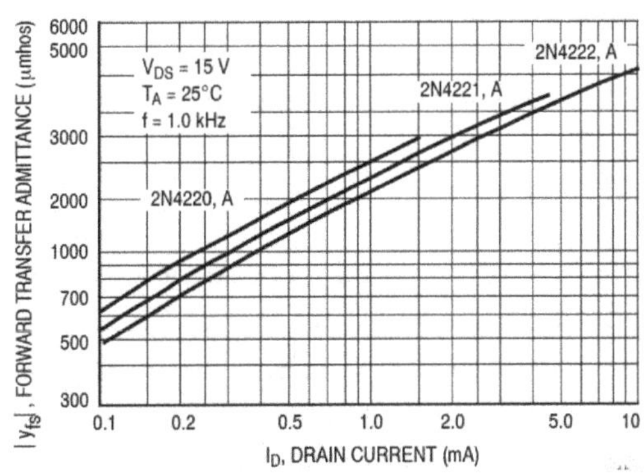

Forward Transfer Admittance versus Drain Current for Typical JFETs

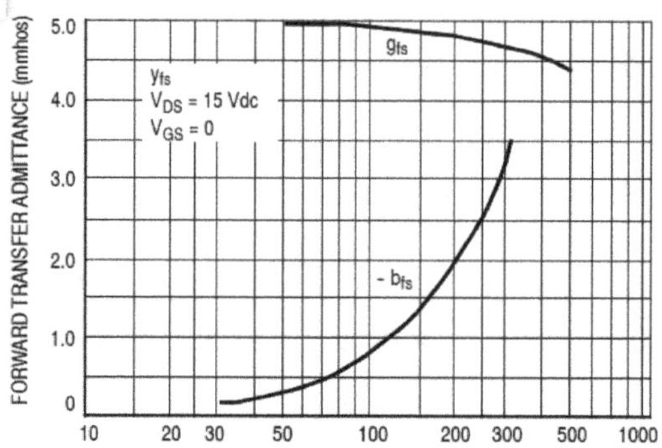

f, FREQUENCY (MHz)
Forward Transfer Admittance versus Frequency

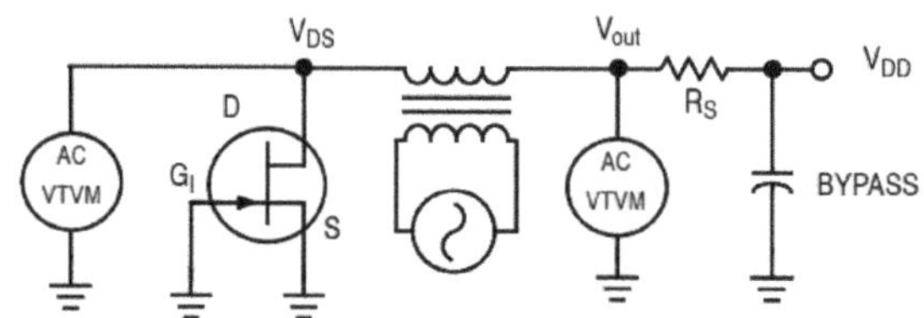

$$Y_{OS} = \frac{V_{out}}{V_{DS}\ R_S}$$

R_S of such value as to cause negligible DC drop.
R_S senses AC drain current

y_{os} Measurement Circuit for Depletion FETs

Voltages and frequencies for measuring y_{os} should be exactly the same as those for measuring y_{fs}. Like y_{fs}, it is a complex number and should be specified as a magnitude at 1 kHz and in complex form at high frequencies.

μ Closely related to y_{os} and y_{fs} is the amplification factor, μ :

$$\mu = \Delta V_{DS} / \Delta V_{GS} \Big|_{I_D = K}$$

The amplification factor does not appear on the field-effect transistor registration format but can be calculated as yfs/yos. For most small-signal applications, m has little circuit significance. It does, however, serve as a general indication of the quality of the field-effect manufacturing process.

C_{iss} The common-source-circuit input capacitance, C_{iss}, takes the place of y_{is} in low-frequency field-effect transistors. This is because y_{is} is entirely capacitive at low frequencies. C_{iss} is conveniently measured in the circuit of Figure for the tetrode JFET. As with y_{fs}, two measurements are necessary for tetrode-connected devices.

At very high frequencies, the real component of y_{is} becomes important so that rf field-effect transistors should have y_{is} specified as a complex number at the same conditions as other high-frequency parameters. For tetrode-connected rf FETs, reading of both Gate 2 to source and Gate 1 tied to Gate 2 are necessary. In switching applications Ciss is of major importance since a large voltage swing at the gate must appear across C_{iss}. Thus, C_{iss} must be charged by the input voltage before turn-on effectively begins.

C_{rss} Reverse transfer admittance (y_{rs}) does not appear on FET data sheets. Instead C_{rss}, the reverse transfer capacitance, is specified at low frequency. Since y_{rs} for a field-effect transistor remains almost completely capacitive and relatively constant over the entire usable FET frequency spectrum, the low-frequency capacitance is an adequate specification. C_{rss} is measured by the circuit of Figure 18. For tetrode FETs, values should be specified for Gate 1 and for both gates tied together.

Again, for switching applications C_{rss} is a critical characteristic. Similar to the Cob of a junction transistor, C_{rss} must be charged and discharged during the switching interval. For a chopper application, C_{rss} is the feed through capacitance for the chopper drive.

$Cd_{(sub)}$ For the MOSFET, the drain-substrate junction capacitance becomes an important characteristic affecting the switching behaviour. $C_{d(sub)}$ appears in parallel with the load in a switching circuit and must be charged and discharged between the two logic levels during the switching interval.

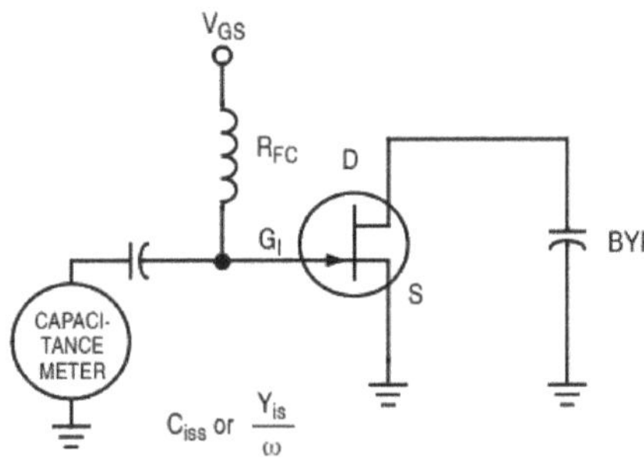

C_{iss} Measurement Circuit

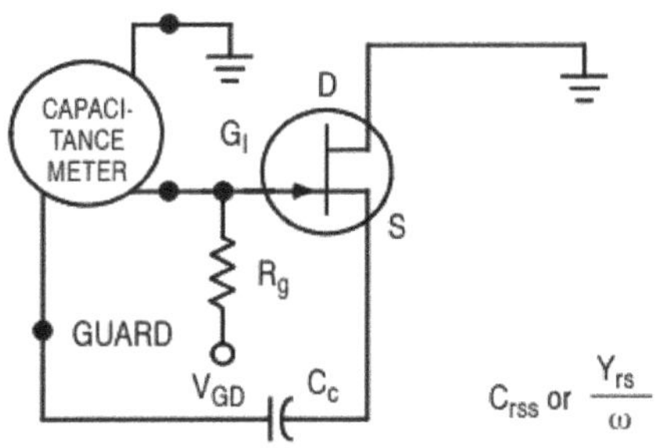

(a) V_{GD} = -- V_{DS}

R_g Typically 1 MΩ, C_c AC Source to guard signal.

Recommended Crss Test Circuit

Noise figure (NF): Like all other active components, field-effect transistors generate a certain amount of noise. The noise figure for field-effect transistors is normally specified on the data sheet as "spot noise", referring to the noise at a particular frequency. The noise figure will vary with frequency and also with the resistance at the input of the device. Typical graphs of such variations are illustrated in Figure for the 2N5458. From graphs of this kind the designer can anticipate the noise level inherent in his design.

rds(on): Channel resistance describes the bulk resistance of the channel in series with the drain and source. From an applications standpoint, it is important primarily for switching and chopper circuits since it affects the switching speed and determines the output level. To complete the confusion of multiple symbols for FET parameters, channel resistance is sometimes indicated as rd(on) and also as r_{DS} and rds. In either case, however, it is measured, for JFETs, by tying the gates to the source, setting all terminals equal to 0 V_{dc}, and applying an ac voltage from drain to source. The magnitude of the ac voltage should be kept low so that there will be no pinchoff in the channel. Insulated-gate FETs may be measured with dc gate bias in the enhancement mode.

R_g, Source Resistance (MEGOHMS) —

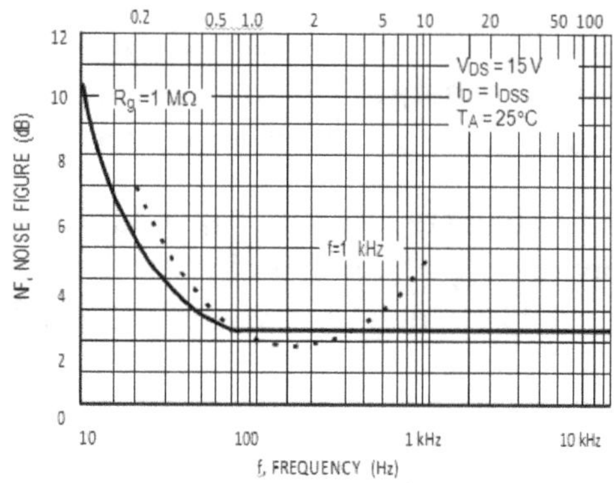

Typical Variations of FET Noise Figure with Frequency and Source Resistance

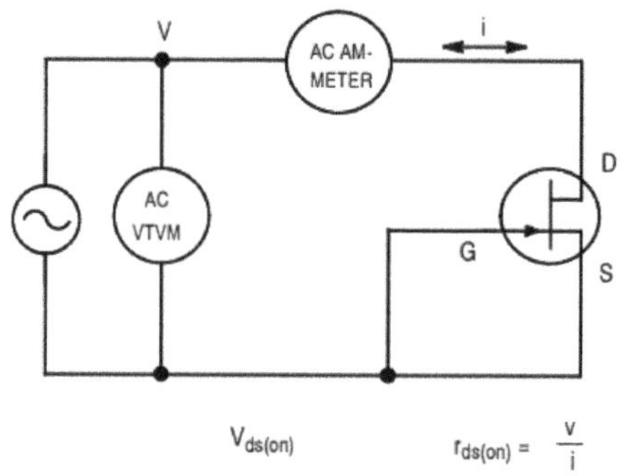

Circuit for Measuring JFET Channel Resistance

Applications

Device Selection

Obviously, different applications call for special emphasis on specific characteristics so that a simple figure of merit that compares devices for all potential uses would be hard to formulate. Nevertheless, an attempt to pinpoint the characteristics that are most significant for various applications has been made* to permit a rapid, first-order evaluation of competitive devices.

The most important single FET parameter, one that applies for any amplifier application, is y_{fs}. This parameter, or one of its many variations, is specified on most data sheets, yet some evaluation is required to come up with a reasonable comparison. For example, in the table of electrical characteristics on most JFET data sheets, y_{fs} is specified at IDSS (V_{GS} = 0) where, for JFETs devices, y_{fs} is maximum. This is illustrated in Figure, where typical variations of y_{fs} as a function of ID are plotted. For some small-signal applications, the IDSS (V_{GS} = 0) point can actually be used as a dc operating point because small-signal excursions into the forward bias region will not actually cause the gate-source junction to become forward-biased. However, in most practical uses, some bias is necessary to allow for the anticipated signal swing; and it must be recognized the y_{fs} goes down as the bias is increased.

It is seen, also, that maximum y_{fs} increases as I_{DSS} increases so that, where maximum y_{fs} is important, a device with a high I_{DSS} specification is normally desirable.

On the other hand, where power dissipation is a factor to be considered, the figure of merit $y_{fs}/V_{GS(off)}$ IDSS has been proposed. This term factors in not only I_{DSS}, which should be low if power dissipation is to be low, but also $V_{GS(off)}$, which indicates maximum input voltage swing.

Since the signal peaks are represented by V_{GS} = VGS(off) and V_{GS} = 0, the lower VGS(off), the higher the figure of merit. And, for amplifier applications requiring a large signal swing, $V_{(BR)GSS}/V_{GS(off)}$ (assuming that VGS(off) is the "pinch-off" voltage) is a satisfactory merit figure because it indicates the ratio of maximum and minimum drain voltages.

For high-frequency circuits, the input capacitance (C_{iss}) and the Miller-effect capacitance (C_{rss}) become important, so $y_{fs}/(C_{iss} + C_{rss})$ indicates a relative measure of device performance. For switching and chopper circuits, a figure of merit is not often useful. Here the magnitudes of Ciss, C_{rrs}, $C_{d(sub)}$ and r_{ds} are of primary interest.

Circuits

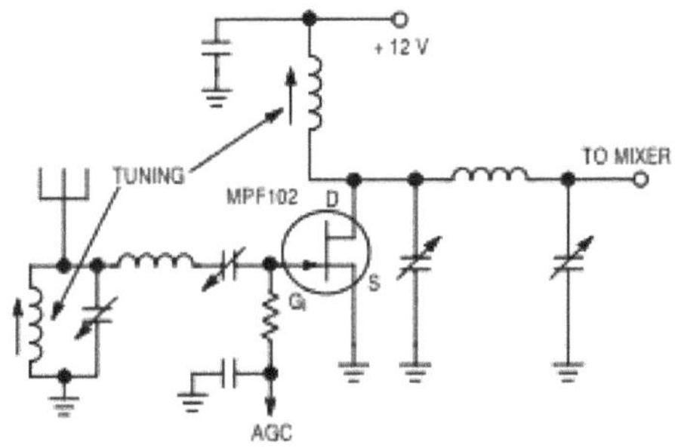

RF Stage of Broadcast Auto Radio

The types of circuits that can utilize FETs are practically unlimited. In fact, many circuits designed to utilize small- signal pentode tubes can utilize FETs with only minor modifications. For example, the circuit in figure shows a typical rf stage for a broadcast-band auto radio. In this circuit, a MPF102 n-channel JFET has replaced the 12BL6 pentode normally employed. The specifications for the two devices, including the AGC characteristics, are similar enough to perform adequately in the circuit of figure.

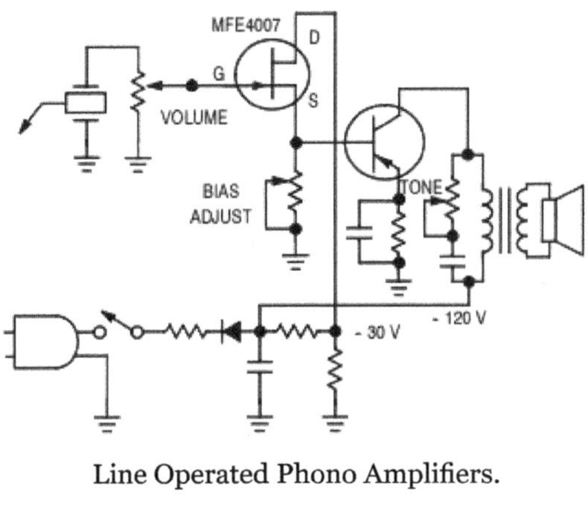

Line Operated Phono Amplifiers.

In an audio application, a field-effect transistor such as the 2N5460 can be combined with a high voltage bipolar transistor to make a simple line-operated phonograph amplifier such as that shown in figure. The ceramic pickup is connected through a potentiometer volume control to the field-effect transistor. Collector current of the transistor, in turn, is set by the potentiometer in the source of the FET. With the proper bipolar output transistor, the circuit can be driven directly from the rectified line voltage, while the low voltage for the FET can be derived from a voltage divider in the power supply line.

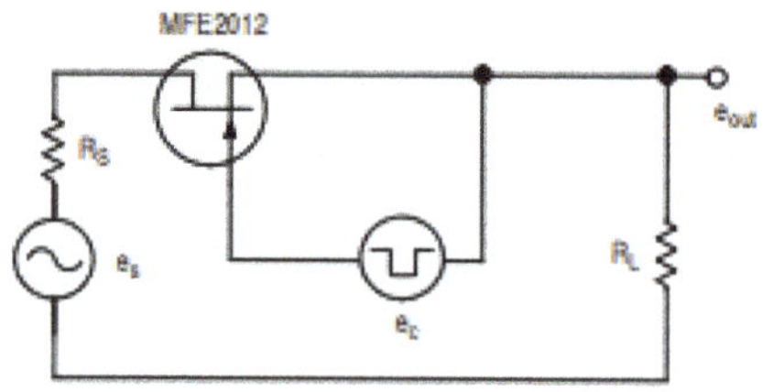

(a) SERIES CHOPPER CIRCUIT

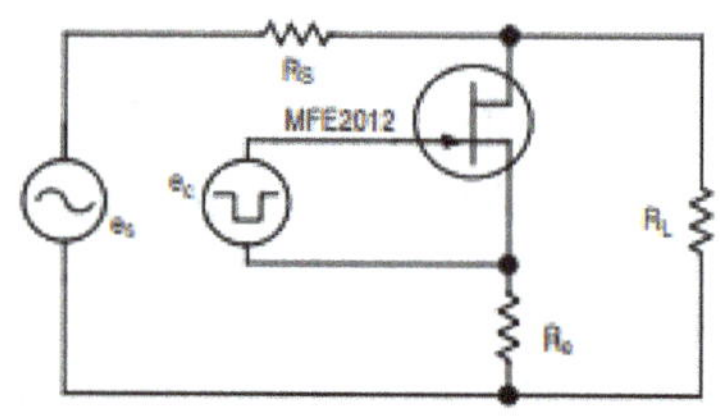

(b) SHUNT CHOPPER

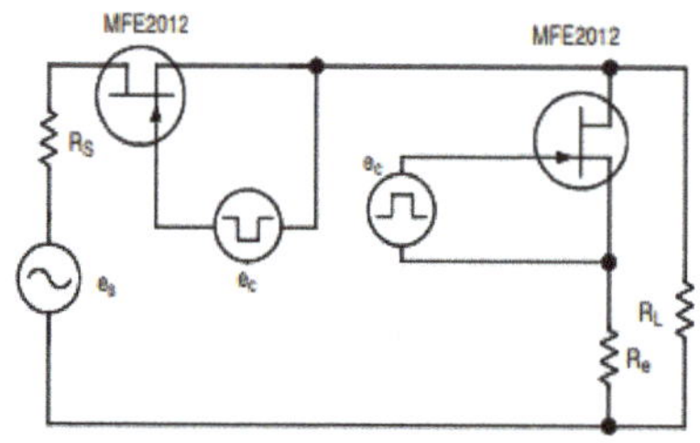

(c) SERIES-SHUNT CHOPPER

Figure above shows three basic chopper circuits. The advantage of the more complex series-shunt circuit is that it balances out the leakage currents of the FETs in order to reduce voltage error and is used to attain high chopping frequencies. From an applications standpoint, the FET circuit is superior to a junction transistor circuit in that there is no offset voltage with the FET turned on. On the minus side, however, the field-effect-transistor chopper generally has a higher series resistance (rds(on)) than the junction transistor.

As newer and better FETs are introduced and as a larger number of designers learn to use them, the range of applications of FETs should broaden considerably.

With its high input impedance, the field-effect transistor will play an important role in input circuitry for instrumentation and audio applications where low-impedance junction transistors have generally been least successful.

Transistor as a Switch

A transistor is used for switching operation for opening or closing of a circuit. This type solid state switching offers significant reliability and lower cost as compared with conventional relays.

Both NPN and PNP transistors can be used as switches. Some of the applications use a power transistor as switching device, at that time it may necessary to use another signal level transistor to drive the high power transistor.

NPN Transistor as a Switch

Based on the voltage applied at the base terminal of a transistor switching operation is performed. When a sufficient voltage ($V_{in} > 0.7$ V)) is applied between the base and emitter, collector to emitter voltage is approximately equal to 0. Therefore, the transistor acts as a short circuit. The collector current V_{cc}/R_c flows through the transistor.

Similarly, when no voltage or zero voltage is applied at the input, transistor operates in cut off region and acts as an open circuit. In this type of switching connection, load (here LED lamp) is connected to the switching output with a reference point. Thus, when the transistor is switched ON, current will flow from source to ground through the load.

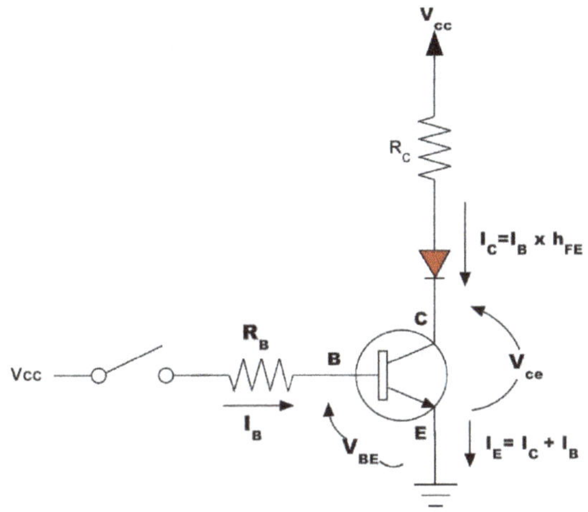

Example of NPN Transistor as a Switch

Consider the below example where base resistance $R_b = 50$k ohm, collector resistance $R_c = 0.7$k ohm, V_{cc} is 5V and the beta value is 125. At the base input a signal varying between 0 and 5V is given so we are going to see the output at the collector by varying the Vi at two states that is 0 and 5V as shown in figure.

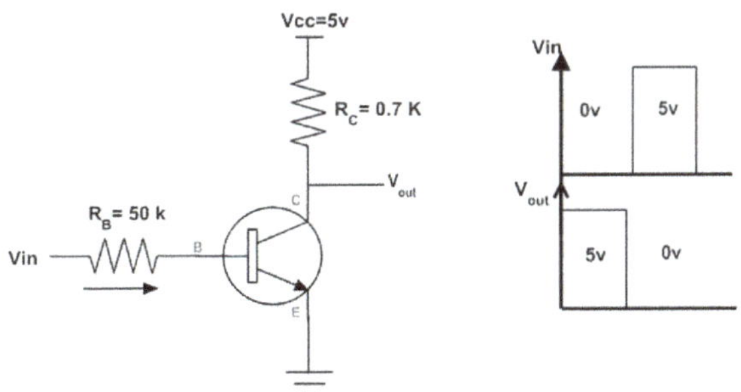

$I_c = V_{cc}/R_c$ when $V_{CE} = 0$

$I_c = 5V/0.7k$ ohm

$I_c = 7.1$ mA

Base Current $I_b = I_c / \beta$

$I_b = 7.1\ mA/125$

$I_b = 56.8\ \mu A$

From the above calculations, the maximum or peak value of the collector current in the circuit is 7.1mA when Vce is equal to zero. And the correspond base current to which collector current flows is 56.8μA. So, it is clear that when the base current is increased beyond the 56.8 micro ampere then the transistor comes into the saturation mode.

Consider the case when zero volt is applied at the input. This causes the base current zero and as the emitter is grounded, emitter base junction is not forward biased. Therefore, the transistor is in OFF condition and the collector output voltage is equal to 5V.

When $V_i = 0V$, $I_b = 0$ and $I_c = 0$,

$$V_c = V_{cc} - (I_c R_c)$$

$$= 5V - 0$$

∞

$$= 5V$$

Consider that input voltage applied is 5 volts, then the base current can be determined by applying Kirchhoff's voltage law.

When, $V_i = 5V$

$$I_b = (V_i - V_{be}) / R_b.$$

For silicon transistor $V_{be} = 0.7\ V$

Thus, $I_b = (5V - 0.7V) / 50K$ ohm

$= 86\ \mu A$ which is greater than $56.8\ \mu A$

Therefore the base current is greater than 56.8 micro ampere current, the transistor will be driven to saturation that is fully ON when 5V is applied at the input. Thus the output at the collector becomes approximately zero.

PNP Transistor as a Switch

PNP transistor works same as NPN for a switching operation, but the current flows from the base. This type of switching is used for negative ground configurations. For the PNP transistor, the base terminal is always negatively biased with respect to the emitter. In this switching, base current flows when the base voltage is more negative. Simply a low voltage or more negative voltage makes transistor to short circuit otherwise it will be open circuited or high impedance state.

In this connection, load is connected to the transistor switching output with a reference point. When the transistor is turned ON, current flows from the source through transistor to the load and finally to the ground.

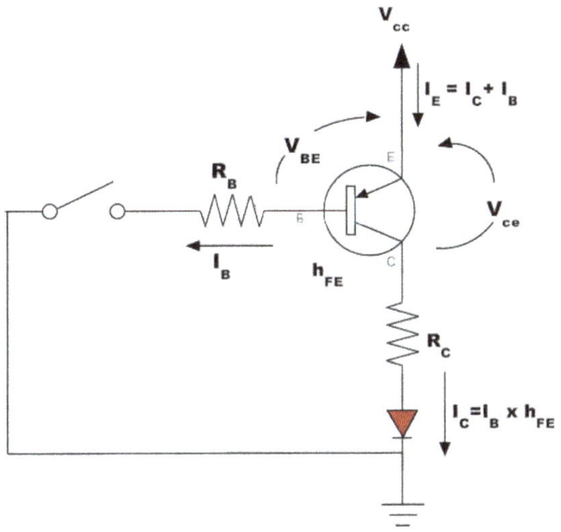

Example of PNP Transistor as a Switch

Similar to the NPN transistor switch circuit, PNP circuit input is also base, but the emitter is connected to constant voltage and the collector is connected to ground through the load as shown in figure.

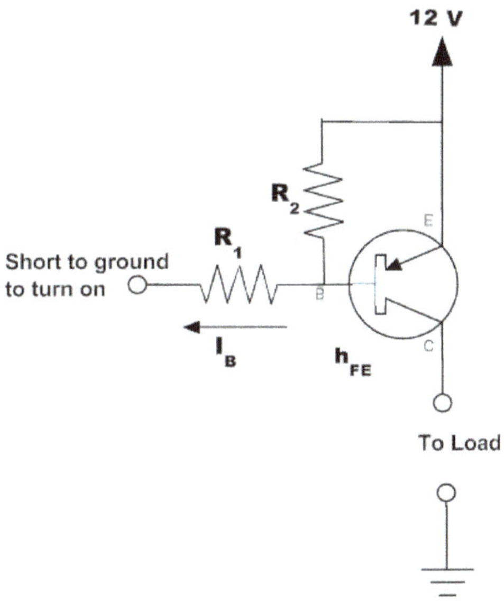

In this configuration base is always biased negatively with respect to the emitter by connecting the base at negative side and the emitter at the positive side of the input supply. So the voltage V_{BE} is negative and the emitter supply voltage with respect to the Collector is positive (V_{CE} positive).

Therefore, for the conduction of transistor emitter must be more positive with respect to both collector and base. In other words base must be more negative with respect to the emitter.

For calculating the base and collector currents following expressions are used.

$$I_c = I_e - I_b$$
$$I_c = \beta . I_b$$
$$I_b = I_c / \beta$$

Consider the above example, that the load requires 100 milli ampere current and the transistor has the beta value of 100. Then the current required for the saturation of the transistor is

$$\text{Minimum base current} = \text{collector current} / \beta$$
$$= 100\text{ mA} / 100$$
$$= 1\text{mA}$$

Therefore, when the base current is 1 mA, the transistor will be fully ON. But practically 30 percent of more current is required for guaranteed saturation of transistor. So, in this example the base current required is 1.3mA.

Common Practical Examples of Transistor as a Switch

Transistor to Switch the LED

The transistor can be used as a switch. The schematic below shows how a transistor is used to switch the Light Emitting Diode (LED).

- When the switch at the base terminal is open, no current flows through the base so the transistor is in the cutoff state. Therefore, the circuit acts as open-circuit and the LED becomes OFF.
- When the switch is closed, base current starts flowing through the transistor and then drives into saturation results to LED become ON.
- Resistors are placed to limit the currents through the base and LED. It is also possible to vary the intensity of LED by varying the resistance in the base current path.

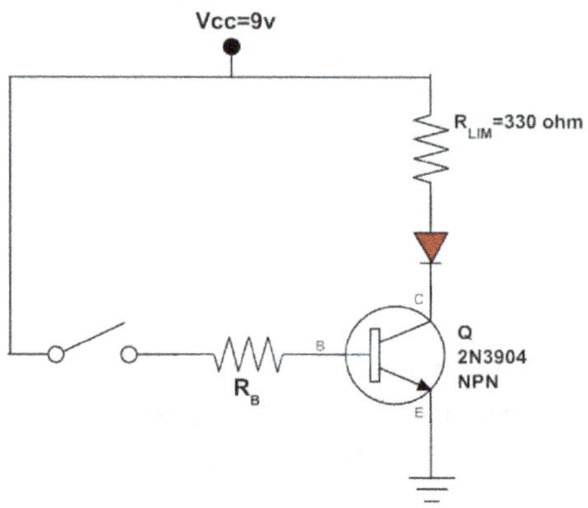

Transistor as an Amplifier

Amplifier circuit can be defined as, a circuit which is used to amplify a signal. The input of the amplifier is a voltage otherwise current, where the output will be an amplifier input signal. An amplifier circuit which uses a transistor otherwise transistors is known as a transistor amplifier. The applications of transistor amplifier circuits mainly involve in audio, radio, optical fiber communication, etc.

The transistor configurations are classified into three types such as CB (common base), CC (common collector), and CE (common emitter). But common emitter configuration is frequently used in the applications like an audio amplifier. Because in CB configuration, the gain is <1, and in CC configuration, the gain is almost equivalent to 1.

The parameters of a good transistor mainly include different parameters namely high gain, high slew rate, high bandwidth, high linearity, high efficiency, high i/p impedance, and high stability etc.

Transistor as an Amplifier Circuit

A transistor can be used as an amplifier by enhancing the weak signal's strength. With the help of the following transistor amplifier circuit, one can get an idea about how the transistor circuit works as an amplifier circuit.

In the below circuit, the input signal can be applied among the emitter-base junction and the output across the Rc load connected in the collector circuit.

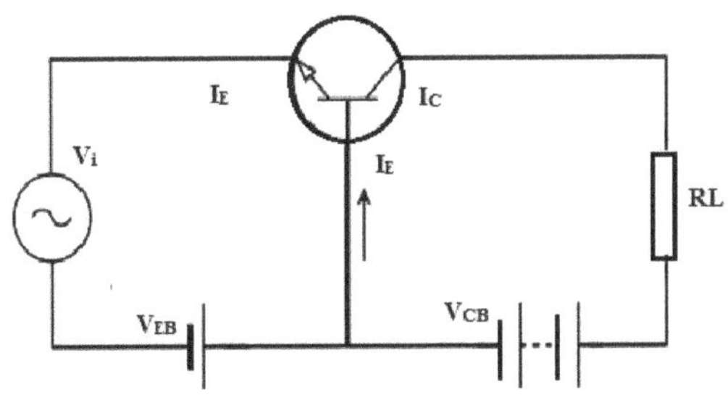

The transistor as an Amplifier Circuit

For accurate amplification, always remember that the input is connected in forward-biased whereas the output is connected in reverse-biased. For this reason, in addition to the signal, we apply DC voltage (VEE) in the input circuit as shown in the above circuit.

Generally, the input circuit includes low resistance as a result; a little change will occur in signal voltage at the input which leads to a significant change within the emitter current. Because of the transistor act, emitter current change will cause the same change within the collector circuit.

Common Base Amplifier

The common base amplifier is the least widely used of the three transistor amplifier configurations. The common emitter and common collector (emitter follower) configurations are far more widely used because their characteristics are generally more useful.

The common base amplifier configuration comes into its own at high frequencies where stability can be an issue.

The emitter follower transistor amplifier characteristics enable the circuit to be used as a buffer amplifier.

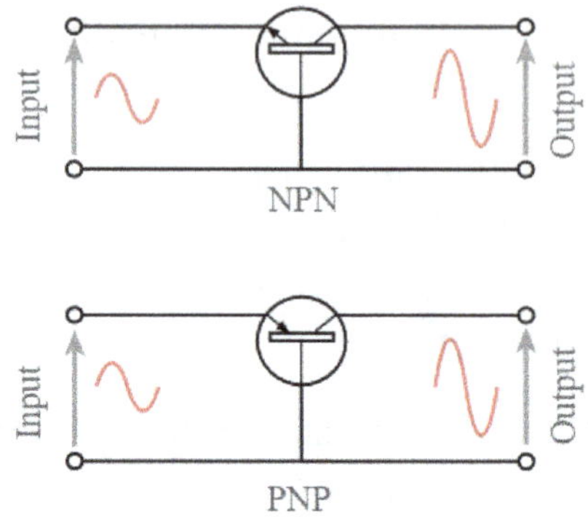

Transistor common base circuit configuration

For both NPN and PNP circuits, it can be seen that for the common base amplifier circuit, the input is applied to the emitter, and the output is taken from the collector. The common terminal for both circuits is the base. The base is grounded for the signal and for this reason the circuit may sometimes be called a grounded base circuit.

The common base amplifier configuration is not used as widely as transistor amplifier configurations. However it does find uses with amplifiers that require low input impedance levels. One application is for moving-coil microphones preamplifiers - these microphones have very low impedance levels.

Another application is within VHF and UHF RF amplifiers where the low input impedance allows accurate matching to the feeder impedance which is typically 50Ω or 75Ω. The configuration also improves stability which is a key issue.

It is worth noting that the current gain of a common-base amplifier is always less than unity.

However the voltage gain is more, but it is a function of input and output resistances (and also the internal resistance of the emitter-base junction). As a result, the voltage gain of a common-base amplifier can be very high.

Common base Transistor Amplifier Characteristics

The table below gives a summary of the major characteristics of the common base transistor amplifier.

Common Base Characteristics	
Parameter	Characteristics
Voltage gain	High
Current gain	Low
Power gain	Low

Input / output phase relationship	0°
Input resistance	Low
Output resistance	High

The common base circuit does not find many applications for low frequency circuits - normally a high input impedance and low output impedance are desirable. However, it finds use in some high-frequency amplifiers, for example for VHF and UHF. In the common base configuration the input capacitance does not suffer from the Miller effect, which degrades the bandwidth of the common-emitter configuration. Also there is a relatively high isolation between the input and output and this means that there is little feedback from the output back to the input, leading to high stability.

Common Base Transistor Amplifier Circuit

The diagram shows how a common base amplifier circuit can be implemented.

The same bias constraints apply to the common base circuit, but the application of the signals is different, allowing the base to be earthed and hence common to both input and output circuits.

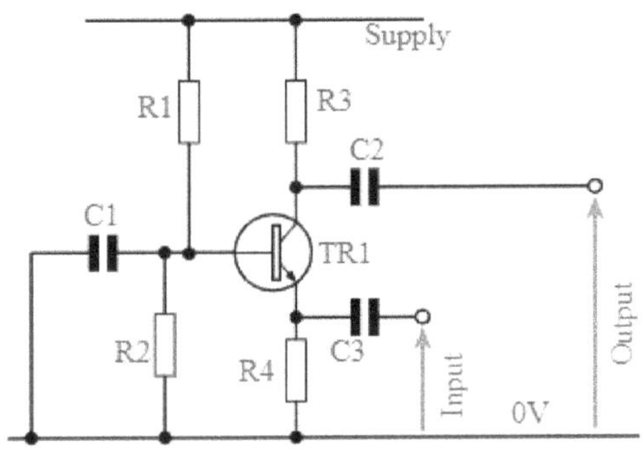

Circuit of a common base transistor amplifier

In this circuit, the same biasing conditions apply. However care has to be taken in the choice of the emitter resistor to ensure that the correct impedance match is provided for the input signal.

Common Collector Amplifier

Common Collector Amplifiers produce an output voltage across its emitter load which is in-phase with the input signal The Common Collector Amplifier is another type of bipolar junction transistor, (BJT) configuration where the input signal is applied to the base terminal and the output signal taken from the emitter terminal. Thus the collector terminal is common to both the input and output circuits. This type of configuration is called Common Collector, (CC) because the collector terminal is effectively "grounded" or "earthed" through the power supply.

In many ways the common collector configuration (CC) is the reverse of the common emitter (CE) configuration as the connected load resistor is changed from the collector terminal for R_C to the emitter terminal for R_E.

The common collector or grounded collector configuration is commonly used where a high impedance input source needs to be connected to a low impedance output load requiring a high current gain. Consider the common collector amplifier circuit below.

Common Collector Amplifier using an NPN Transistor

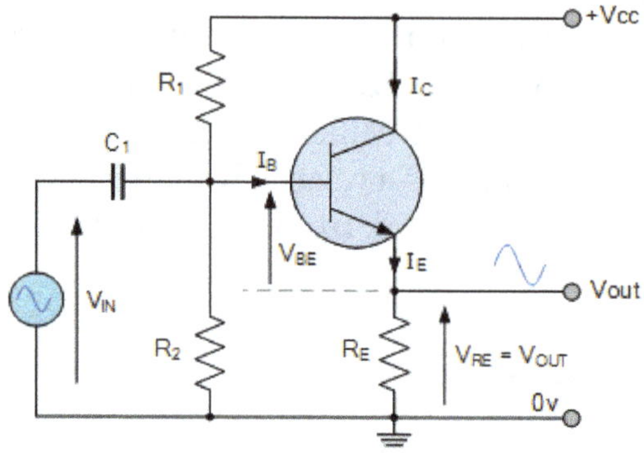

Resistors R_1 and R_2 form a simple voltage divider network used to bias the NPN transistor into conduction. Since this voltage divider lightly loads the transistor, the base voltage, V_B can be easily calculated by using the simple voltage divider formula as shown.

Voltage Divider Network

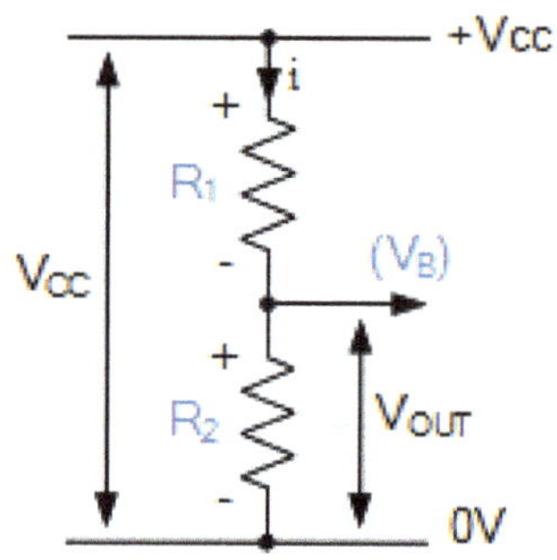

Voltage Divider Equation:

$$V_{OUT} = V_{CC}\left(\frac{R_2}{R_1 + R_2}\right)$$

Therefore,

$$V_B = V_{CC}\left(\frac{R_2}{R_1 + R_2}\right)$$

With the collector terminal of the transistor connected directly to V_{CC} and no collector resistance, (R_C = 0) any collector current will generate a voltage drop across the emitter resistor R_E.

However, in the common collector amplifier circuit, the same voltage drop, V_E also represents the output voltage, V_{OUT}.

Ideally we would want the DC voltage drop across R_E to be equal to half the supply voltage, V_{CC} to make the transistors quiescent output voltage sit somewhere in the middle of the characteristics curves allowing for a maximum unclipped output signal. Thus the choice of R_E depends greatly on I_B and the transistors current gain Beta, β.

As the base-emitter pn-junction is forward biased, base current flows through the junction to the emitter encouraging transistor action causing a much larger collector current, I_C to flow. Thus the emitter current is a combination of base current and collector current as: $I_E = I_B + I_C$. However, as the base current is extremely small compared to the collector current, the emitter current is therefore approximately equal to the collector current. Thus $I_E \approx I_C$ as with the common emitter (CE) amplifier configuration, the input signal is applied to the transistors base terminal, the amplifiers output signal is taken from the emitter emitter terminal. However, as there is only one forward biased pn-junction between the transistors base and its emitter terminal, any input signal applied to the base passes directly through the junction to the emitter. Therefore the output signal present at the emitter is in-phase with the applied input signal at the base.

As the amplifiers output signal is taken from across the emitter load this type of transistor configuration is also known as an Emitter Follower circuit as the emitter output "follows" or tracks any voltage changes to the base input signal, except that it remains about 0.7 volts (VBE) below the base voltage. Thus V_{IN} and V_{OUT} are in-phase producing zero phase difference between the input and output signals.

The emitters pn-junction effectively acts as a forward biased diode and for small AC input signals this emitter diode junction has a resistance given by: $r'_e = 25mV/I_e$ where the 25mV is the thermal voltage of the junction at room temperature (25 °C) and Ie is the emitter current. So as the emitter current increases, the emitter resistance decreases by a proportional amount.

The base current which flows through this internal base-emitter junction resistance also flows out and through the externally connected emitter resistor, R_E. These two resistances are series connected thus acting as a potential divider network creating a voltage drop. Since the value of r'e is very small, and RE is much larger, usually in the kilohms (kΩ) range, the magnitude of the amplifiers output voltage is therefore less than its input voltage.

However, in reality the magnitude of the output voltage (peak-to-peak) is generally in the 98 to 99% value of the input voltage which is close enough in most cases to be considered as unity gain.

We can calculate the voltage gain, V_A of the common collector amplifier by using the voltage divider formula as shown assuming that the base voltage, V_B is actually the input voltage, V_{IN}.

Common Collector Amplifier Voltage Gain

$$V_{OUT} = \frac{V_{IN} \times R_E}{r^1_e + R_E}$$

Thus,

$$A_V = \frac{V_{OUT}}{V_{IN}} = \frac{I_e \times R_E}{Ie(r^1_e + R_E)}$$

Since, R_E is much greater than $r^1_e \, (r^1_e + R_E) = R_E$

And the two emitter currents I_e cancel thus:

$$A_V = \frac{V_{OUT}}{V_{IN}} = \frac{R_E}{R_E} = 1$$

So the common collector amplifier cannot provide voltage amplification and another expression used to describe the common collector amplifier circuit is as a Voltage Follower Circuit for obvious reasons. Thus since the output signal closely follows the input and is in-phase with the input the common collector circuit is therefore a non-inverting unity voltage gain amplifier.

Common Collector Amplifier

Example: A common collector amplifier is constructed using an NPN bipolar transistor and a voltage divider biasing network. If R_1 = 5k6Ω, R_2 = 6k8Ω and the supply voltage is 12 volts. Calculate the values of: V_B, V_C and V_E, the emitter current IE, the internal emitter resistance r'e and the amplifiers voltage gain A_V when a load resistance of 4k7Ω is used. Also draw the final circuit and corresponding characteristics curve with load line.

1. Base biasing voltage, V_B

$$I = \frac{V_{CC}}{R_1 + R_2} = \frac{12}{5600 + 6800} = 968\,\mu A$$

$$V_B = I \times R_2 = 968 \times 10^{-6} \times 6800 = 6.5V$$

2. Collector voltage, V_C. As there is no collector load resistance, the transistors collector terminal is connected directly to the DC supply rail, so $V_C = V_{CC}$ = 12 volts.

3. Emitter biasing voltage, V_E

$$V_E = V_B - V_{BE} = 6.5 - 0.7 = 5.8V$$

Thus:

$$V_{CE(OFF)} = V_{CC} - V_E = 12 - 5.8 = 6.2V$$

4. Emitter Current, I_E

$$I_E = \frac{V_E}{R_E} = \frac{5.8}{4700} = 0.00123 = 1.23mA$$

5. AC Emitter Resistance, r'_e

$$r^1_e = \frac{25mV}{I_E} = \frac{25mV}{1.23mA} = 20.3\Omega$$

6. Voltage gain, A_V

$$A_V = \frac{R_E}{r_e + R_E} = \frac{4700}{20.3 + 4700} = 0.996 \text{ Or } 99.6\%$$

Common Collector Amplifier Circuit with Load Line

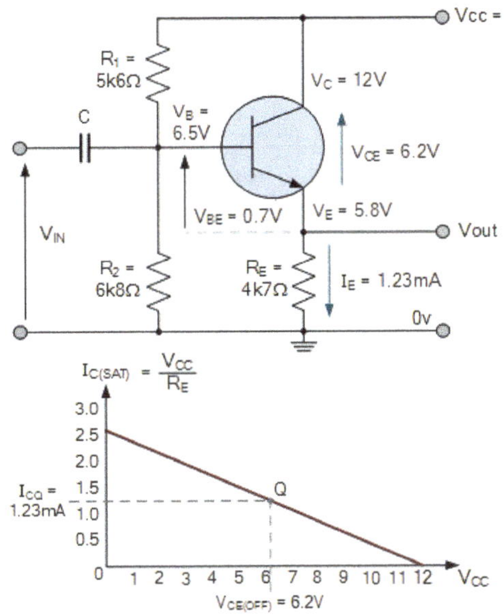

Common Collector Input Impedance

Although the common collector amplifier is not very good at being a voltage amplifier, because as we have seen, its small signal voltage gain is approximately equal to one (AV ≅ 1), it does however make a very good voltage buffer circuit due to its high input (ZIN) and low output (ZOUT) impedances, providing isolation between an input signal source from a load impedance load.

Another useful feature of the common collector amplifier is that it provides current gain (Ai) as long as it is conducting. That is it can pass a large current flowing from the collector to the emitter, in response to a small change to its base current, I_B. Remember that this DC current only sees R_E as there is no R_C. Then the DC current is simply: V_{CC}/R_E which can be large if RE is small.

Consider the basic common collector amplifier or emitter follower configuration below.

Common Collector Amplifier Configuration

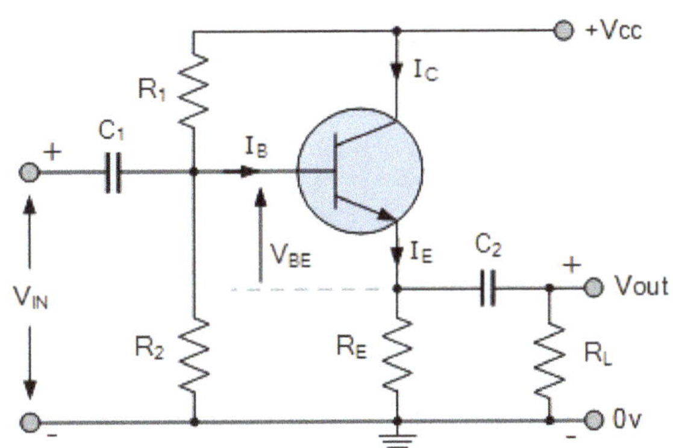

For AC analysis of the circuit, the capacitors are shorted and V_{CC} is shorted (zero impedance). Thus the equivalent circuit is given as shown with the biasing currents and voltages given as:

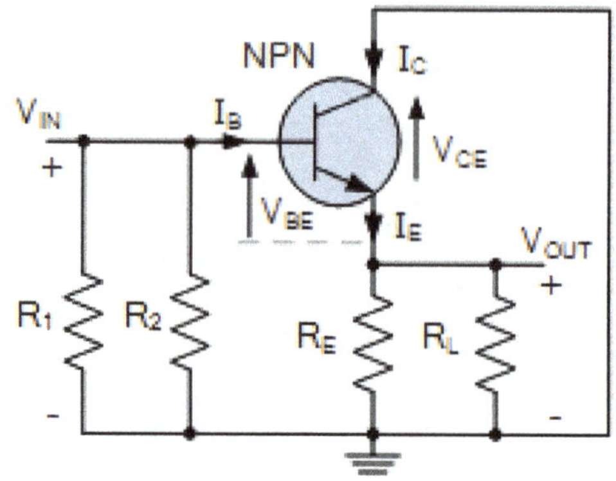

$$R_B = R_1 \| R_2$$

$$R_e = R_e \| R_L$$

$$\beta = \frac{I_C}{I_B} \therefore I_C = \beta I_B$$

$$I_E \approx I_C = \beta I_B$$

$$V_{IN} = V_B = V_{BE} + V_E$$

The Input Impedance, Z_{IN} of the common collector configuration looking into the base is given as:

$$Z_{IN} = R_{BIAS} \| Z_{base} \text{ where}: Z_{base} = \frac{V_b}{i_b}$$

$$\text{But}: V_b = i_e (R_e + r'_e), i_e = i_C + i_b \text{ and } i_c = \beta i_o$$

$$\text{so}: Z_{base} = \frac{V_b}{i_b} = \frac{i_e(R_e + r^1 e)}{i_b} = \frac{(i_c + i_b)(R_e + r^1 e)}{i_b}$$

$$= \frac{(\beta i_b + i_b^1)(R_e + r^1_e)}{i_b} = (\beta + 1)(R_e + r^1{}_e)$$

But as Beta, β is generally much greater than 1 (usually above 100), the expression of: β + 1 can be reduced to just Beta, β as multiplication by 100 is virtually the same as multiplying by 101. Thus -

Common Collector Amplifier Base Impedance

$$Z_{base} = \beta(R_e + r'_e)$$

Where: β is the transistors current gain, Re is the equivalent emitter resistance, and r'e is the ac resistance of the emitter-base diode. Note that since the combined value of Re is generally much greater than the diodes equivalent resistance, r'e (kilo-ohms compared to a few ohms) the transistors base impedance can be given as simply: $\beta * R_e$.

An interesting point to notice here is that the the transistors input base impedance, $Z_{IN(base)}$ can be controlled by the value of either the emitter leg resistor, RE or the load resistor R_L as they are parallel connected.

While the equation above gives us the input impedance looking into the base of the transistor, it does not give us the true input impedance that the source signal would see looking into the complete amplifier circuit. For that we need to consider the two resistors which make up the voltage divider biasing network. Thus-

Common Collector Amplifier Input Impedance

$$Z_{IN} = R_{BIAS} \| Z_{base}$$

$$\text{where}: R_{BIAS} = R_B = R_1 \| R_2$$

$$\therefore Z_{IN} = R_B \| \beta(R_e + r'_e)$$

Common Collector

Example: Using the previous common collector amplifier circuit above, calculate the input impedances of the transistors base and amplifier stage if the load resistance, RL is 10kΩ and the NPN transistors current gain is 100.

1. AC Emitter Resistance, r'_e

$$r'_e = \frac{25\,\text{mV}}{I_E} = \frac{25\,\text{mV}}{1.23\,\text{mA}} = 20.3\Omega$$

2. Equivalent Load Resistance, R_e

$$R_e = R_E \| R_L$$

$$R_e = \frac{R_E \times R_L}{R_E + R_L} = \frac{4700 \times 10000}{4700 + 10000} = 3197\Omega \text{ or } 3.2\,\text{k}\Omega$$

3. Transistors Base Impedance, Z_{BASE}

$$Z_{BASE} = \beta(R_e + r'_e)$$

$$r'_e = \frac{25\,\text{mV}}{I_E} = \frac{25\,\text{mV}}{1.23\,\text{mA}} = 20.3\Omega$$

$$\therefore Z_{BASE} = 100(3200 + 20.3) = 322\,\text{k}\Omega$$

4. Amplifier Input Impedance, $ZIN_{(STAGE)}$

$$Z_{IN} = R_{BIAS} \| Z_{BASE}$$

$$R_{BIAS} = \frac{R_1 \times R_2}{R_1 + R_2} = \frac{5600 \times 6800}{5600 + 6800} = 3070\Omega$$

$$\therefore Z_{IN} = \frac{R_{BIAS} \times Z_{BASE}}{R_{BIAS} + Z_{BASE}} = \frac{3070 \times 32000}{3070 + 32000} = 2800\Omega \text{ or } 2.8\,k\,\Omega$$

As the transistors base impedance of 322kΩ is much higher than the amplifiers input impedance of only 2.8kΩ, thus the input impedance of the common collector amplifier is determined by the ratio of the two biasing resistors, R_1 and R_2.

Common Collector Output Impedance

To determine the CC amplifiers output impedance Z_{OUT} looking from the load back into the amplifiers emitter terminal, we must first remove the load as we want to see the effective resistance of the amplifier that is driving the load. Thus the AC equivalent circuit looking into the amplifiers output is given as:

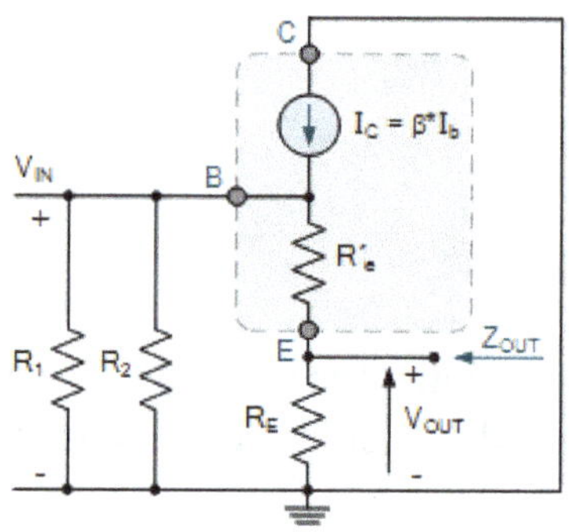

From above, the input impedance of the base circuit is given as: $R_B = R_1||R_2$. The current gain of the transistor is given as: β. Thus the output equation is given as:

$$Z_{OUT} = R_E \| \left(r_e' + \frac{R_1 \| R_2}{\beta + 1} \right)$$

We can see then that the emitter resistor, RE is effectively in parallel with the whole impedance of the transistor looking back into its emitter terminal.

If we calculate the output impedance of our common emitter amplifier circuit using the component values from above, it would give an output impedance ZOUT of less than 50Ω (49.5Ω) which is much smaller than the higher input impedance, $ZIN_{(BASE)}$ calculated previously.

Thus we can see then that the Common Collector Amplifier configuration has, from calculation, a very high input impedance and a very low output impedance allowing it to drive a low impedance load. In fact due to the CC amplifiers relatively high input impedance and very low output impedance it is commonly used as a unity gain buffer amplifier.

Having determined that the output impedance, Z_{OUT} of our example amplifier above is approximately 50Ω by calculation, if we now connect the 10kΩ load resistor back into the circuit, the resulting output impedance will be:

$$Z_{OUT(LOAD)} = Z_{OUT} \parallel R_L$$

$$Z_{OUT(LOAD)} = \frac{49.5 \times 10000}{49.5 + 10000} = 49.3\Omega$$

Although the load resistance is 10kΩ, the equivalent output resistance is still low at 49.3Ω. This is because R_L is large compared with Z_{OUT}, thus for maximum power transfer, RL must equal Z_{OUT}. As the voltage gain of the common collector amplifier is considered to be unity (1), the amplifiers power gain must be equal to its current gain, as P = V*I.

Since the common collector current gain is defined as the ratio of the emitter current to the base current, $\gamma = I_E/I_B = \beta + 1$, it therefore follows that the amplifiers current gain must be approximately equal to Beta (β) as β + 1 is virtually the same as Beta.

Common Emitter Amplifier

The common emitter amplifier is a three basic single stage bipolar junction transistor and is used as a voltage amplifier. The input of this amplifier is taken from the base terminal, the output is collected from the collector terminal and the emitter terminal is common for both the terminals. The basic symbol of the common emitter amplifier is shown below.

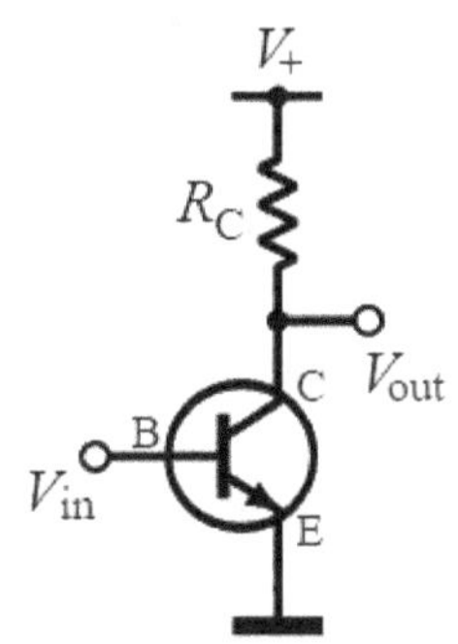

Common Emitter Amplifier

Working of Common Emitter Amplifier

The below circuit diagram shows the working of the common emitter amplifier circuit and it consists of voltage divider biasing, used to supply the base bias voltage as per the necessity. The voltage divider biasing has a potential divider with two resistors are connected in a way that the midpoint is used for supplying base bias voltage.

There are different types of electronic components in the common emitter amplifier which are R1 resistor is used for the forward bias, the R2 resistor is used for the development of bias, the RL resistor is used at the output it is called as the load resistance. The RE resistor is used for the thermal stability. The C1 capacitor is used to separate the AC signals from the DC biasing voltage and the capacitor is known as the coupling capacitor.

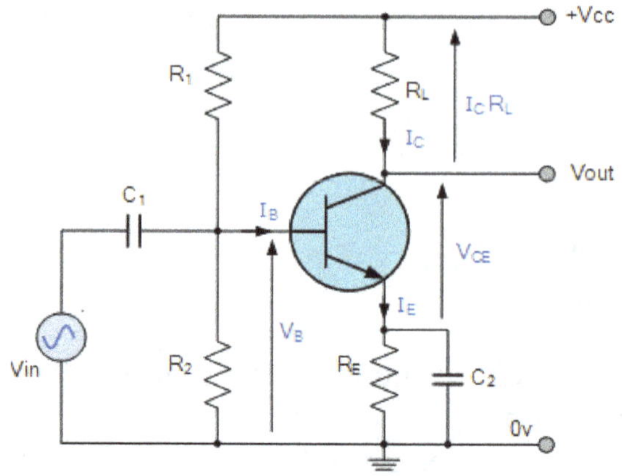

Common Emitter Amplifier Circuit

The figure shows that the bias vs gain common emitter amplifier transistor characteristics, if the R2 resistor increases then there is an increase in the forward bias and R1 & bias are inversely proportional to each other. The alternating current is applied to the base of the transistor of the common emitter amplifier circuit then there is a flow of small base current. Hence there is a large amount of current flow through the collector with the help of the RC resistance. The voltage near the resistance RC will change because the value is very high and the values are from the 4 to 10kohm. Hence there is a huge amount of current present in the collector circuit which amplified from the weak signal, therefore common emitter transistor work as an amplifier circuit.

Voltage Gain of Common Emitter Amplifier

The current gain of common emitter amplifier is defined as the ratio of change in collector current to the change in base current. The voltage gain is defined as the product of the current gain and the ratio of the output resistance of the collector to the input resistance of the base circuits. The following equations show the mathematical expression of the voltage gain and the current gain.

$$\beta = \frac{\Delta/c}{\Delta/b}$$

$$Av = \beta \frac{Rc}{Rb}$$

Voltage & Current Gain of Common Emitter Amplifier

Common Emitter Transistor Amplifier

The following circuit diagram shows the common emitter transistor amplifier. This transistor amplifier has a common configuration and it is a standard format of transistor circuit whereas voltage gain is desired. The common emitter amplifier is also converted as an inverting amplifier. The different types of configurations in transistor amplifiers are common base and the common collector transistor and the figure are shown in the following circuits.

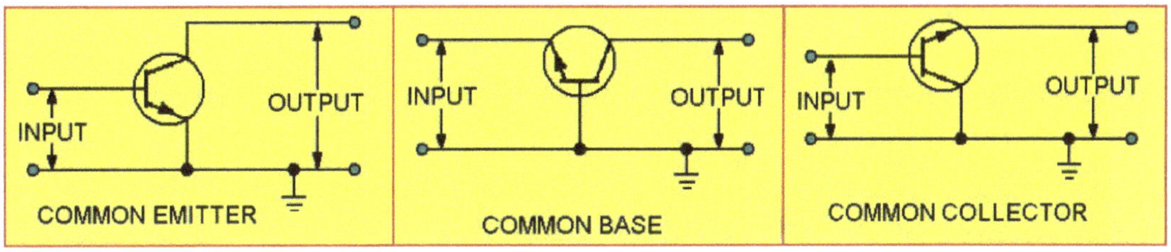

Common Emitter Transistor Amplifier

Characteristics of Common Emitter Amplifier

- The voltage gain of common emitter amplifier is medium,
- The power gain is high in the common emitter amplifier,
- There is a phase relationship of 180 degrees in input and output,
- In the common emitter amplifier, the input and output resistors are medium.

The characteristics graph between the bias and the gain is shown below.

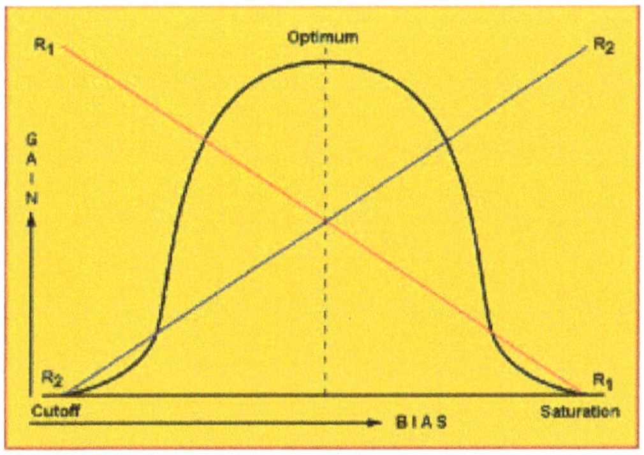

Characteristics Graph

Applications of Common Emitter Amplifier

- The common emitter amplifiers are used in the low-frequency voltage amplifiers.
- These amplifiers are used typically in the RF circuits.
- In general, the amplifiers are used in the Low noise amplifiers.

Advantages of Common Emitter Amplifier

- The common emitter amplifier has a low input impedance and it is an inverting amplifier.
- The output impedance of this amplifier is high.
- This amplifier has highest power gain when combined with medium voltage and current gain.
- The current gain of the common emitter amplifier is high.

Disadvantages of Common Emitter Amplifier

- In the high frequencies, the common emitter amplifier does not respond.
- The voltage gain of this amplifier is unstable.
- The output resistance is very high in these amplifiers.
- In these amplifiers, there is a high thermal instability.
- High output resistance.

References

- Transistor: circuitglobe.com, Retrieved 26 May, 2019
- Transistor: electronics-tutorials.ws, Retrieved 16 February, 2019
- Npn-transistor: electronicshub.org, Retrieved 18 March, 2019
- Pnp-transistor: electronicshub.org, Retrieved 9 July, 2019
- What-is-a-fet-types, fet-field-effect-transistor, electronic_components: electronics-notes.com, Retrieved 12 June, 2019
- Transistor_as_a_Switch: electronicshub.org, Retrieved 19 March, 2019
- Transistor-as-an-amplifier-circuit-diagram-and-its-working: elprocus.com, Retrieved 26 July, 2019
- Transistor-common-base-circuit, transistor, analogue_circuits: electronics-notes.com, Retrieved 28 August, 2019
- Common-collector-amplifier, amplifier: electronics-tutorials.ws, Retrieved 3 January, 2019
- Common-emitter-amplifier-circuit-working: elprocus.com, Retrieved 6 April, 2019

4

Electronic Circuits

The electronic circuit is a circuit that is made up of individual electronic components that are connected by conductive wires through which electric current can flow. Analog circuits and digital circuits are the two primary types of electronic circuits used across all electronic devices. The chapter closely examines the key concepts of electronic circuits to provide an extensive understanding of the subject.

An electronic circuit is a complete course of conductors through which current can travel. Circuits provide a path for current to flow. To be a circuit, this path must start and end at the same point. In other words, a circuit must form a loop. An electronic circuit and an electrical circuit has the same definition, but electronic circuits tend to be low voltage circuits.

For example, a simple circuit may include two components: a battery and a lamp. The circuit allows current to flow from the battery to the lamp, through the lamp, then back to the battery. Thus, the circuit forms a complete loop.

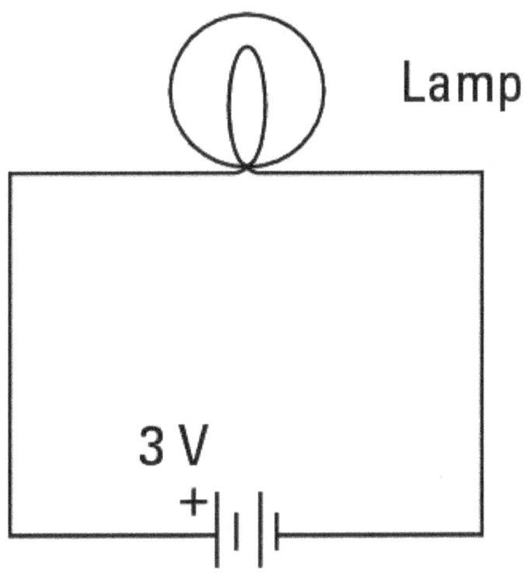

Circuits can be more complex. However, all circuits can be distilled down to three basic elements:

- Voltage source: A voltage source causes current to flow like a battery, for instance.
- Load: The load consumes power; it represents the actual work done by the circuit. Without the load, there's not much point in having a circuit. The load can be as simple as a single light bulb. In complex circuits, the load is a combination of components, such as resistors, capacitors, transistors, and so on.
- Conductive path: The conductive path provides a route through which current flows. This route begins at the voltage source, travels through the load, and then returns to the voltage source. This path must form a loop from the negative side of the voltage source to the positive side of the voltage source.

The following paragraphs describe a few additional interesting points to keep in mind as you ponder the nature of basic circuits:

- When a circuit is complete and forms a loop that allows current to flow, the circuit is called a closed circuit. If any part of the circuit is disconnected or disrupted so that a loop is not formed, current cannot flow. In that case, the circuit is called an open circuit.
- Short circuit refers to a circuit that does not have a load. For example, if the lamp is connected to the circuit but a direct connection is present between the battery's negative terminal and its positive terminal, too.

Analog Circuit

An Analog Circuit is a type of Electronic Circuit which processes analog data using analog components like resistors, capacitors, diodes, transistors etc.

Analog Circuits can be quite simple like a combination of resistors to form a voltage divider or a combination of Op-amps (which internally contain transistors), resistors, diodes etc. to form an amplifier.

The following is an example of a simple Analog Circuit.

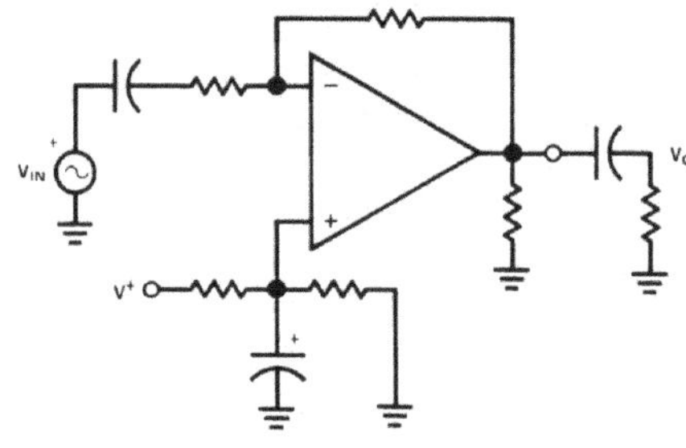

This is a simple amplifier circuit implemented using Op-amp, resistors and capacitors. By selecting appropriate values for resistors and capacitors, you can achieve a significant gain at the output.

Cathode Ray Oscilloscope (CRO) is an Analog Device that can detect ad analyse analog signals.

Analog circuits represent key components of communications and other systems in widespread, growing commercial use. High-speed transistors are essential to the operation of such circuits. The modest complexity of the analog circuits (e.g. hundreds of devices), and the relaxed requirements on on/off ratio and other aspects, make SWNT array transistors well suited to this application area, especially novel flexible analog devices. RF transistors based on SWNT arrays have been demonstrated at frequencies up to 10 GHz, with relatively coarse device dimensions. They can therefore be further configured as RF power amplifiers and oscillators, two of the most important building blocks of analog electronic systems. As an example, these devices can then be connected and impedance matched to generate functional radio receiving systems, where the SWNT TFTs provide all of the active components, even including the audio output amplifier stage. An optical image of a completed SWNT radio is shown in figure and the SWNT chips are connected to an external antenna and a speaker through wire bonding. This relatively simple system, and others like it, is fully functional and capable of receiving, demodulating and amplifying signals broadcast by commercial radio stations. Figure presents the power spectrum recorded during a weather/traffic report that is directly audible through headphones connected to the system.

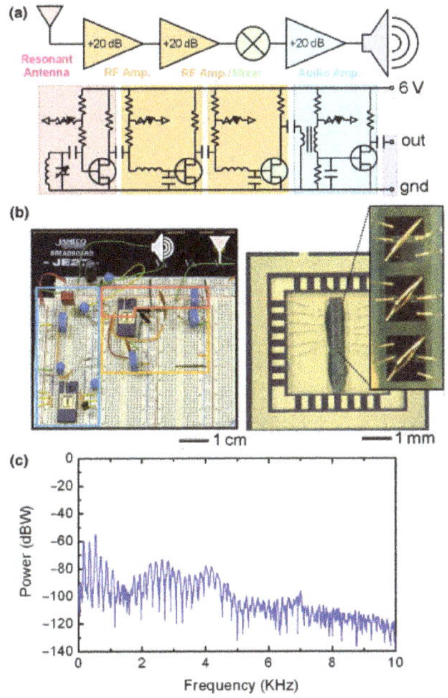

Analog circuit elements support a continuous range of voltages, and the information they carry is usually encoded as a time-varying continuous signal. All of the circuits described thus far are analog circuits. Analog circuitry is a necessary part of most medical instrumentation because the devices that perform physiological measurements, the transducers or “biotransducers,” usually produce analog electric signals. This includes devices that measure movement, pressure, bioelectric activity, sound and ultrasound, light, and other forms of electromagnetic energy. Bioelectric signals such as the EEG, ECG, and EMG are analog signals. Before these analog signals can be processed by a digital computer, some type of manipulation is usually required while the signals are still in the analog domain. This analog signal processing may only consist of increasing the amplitude of the signal, but can also include filtering and other basic signal-processing operations. Unlike digital circuitry, the design of analog circuits is often the responsibility of the biomedical engineer. After analog signal processing, the signal is usually converted to a digital signal using an analog-to-digital converter. The components of a typical biomedical instrument are summarized in figure.

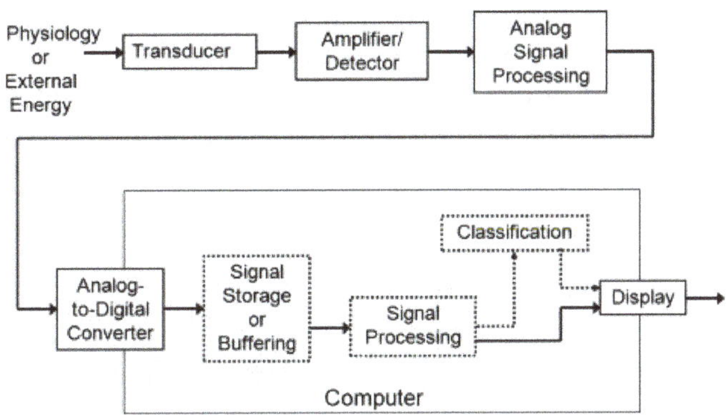

Analogue circuits which deliberately generate high frequency signals (remembering that the emissions regulatory regime currently begins at 150kHz, and in some cases lower than this) need to follow the same layout, decoupling and grounding rules as already outlined. It is also possible for low frequency analogue circuits to operate unintentionally outside their design bandwidth.

Instability

Analogue amplifier circuits may oscillate in the MHz region and thereby cause interference for a number of reasons:

- Feedback-loop instability;
- Poor decoupling and
- Output stage instability.

Capacitive coupling due to poor layout and common-impedance coupling are also sources of oscillation. Any prototype amplifier circuit should be checked for HF instability, whatever its nominal bandwidth, in its final configuration and layout. Feedback instability is due to too much feedback near the unity-gain frequency, where the amplifier's phase margin is approaching a critical value. It may be linked with incorrect compensation of an uncompensated op-amp.

Even if the circuit doesn't actually oscillate, near-instability is a potential problem for RF immunity as well: if the amplifier circuit is configured so that its stability margin is reduced at frequencies of a few MHz, then RF coupled into the circuit at around these frequencies will be much more likely to cause susceptibility effects.

Decoupling

Power supply rejection ratio falls with increasing frequency, and power supply coupling to the input at high frequencies can be significant in wideband circuits. The higher the impedance of the power rails, the more unwanted coupling exists.

This is cured by good decoupling at the power supply pins, but typical 0.01–0.1μF decoupling capacitors may resonate with the parasitic inductance of long power leads in the MHz region, so decoupling-related instability problems may show up in this range. Paralleling a low-value capacitor with a 1–10μF electrolytic capacitor will drop the resonant frequency and stray circuit Q to a manageable level. The electrolytic's series inductance could resonate with the ceramic capacitor and actually worsen the situation, although the losses of a few ohms in series with the capacitor usually prevent this, which is a good argument for not using low-loss 1–10μF ceramic capacitors in this case. The input stages of multi-stage high gain amplifiers may need additional resistance or a ferrite bead in series with each stage's supply to improve decoupling from the power rails.

Output Stage Instability

Capacitive loads cause a phase lag in the output voltage by acting in combination with the op-amp's open-loop output resistance. This increased phase shift reduces the phase margin of a feedback circuit, possibly by enough to cause oscillation. A typical capacitive load, often invisible to the designer because it is not treated as a component, is a length of unterminated coaxial cable. Until the length starts to approach a quarter-wavelength at the frequency of interest, coax looks like a capacitor: for instance, 10m of the popular RG58C/U 50Ω type will be about 1000pF. Other capacitive loads may be found within a system, such as a sample-and-hold circuit, but the cable load is of special interest for EMC since it couples any RF oscillations both into and out of the circuit. To cure output instability, decouple the capacitance from the output with a low-value series resistor, and add high frequency feedback with a small direct feedback capacitor CF which compensates for the phase lag caused by CL. When the critical frequency is high a ferrite chip is an acceptable substitute for RS.

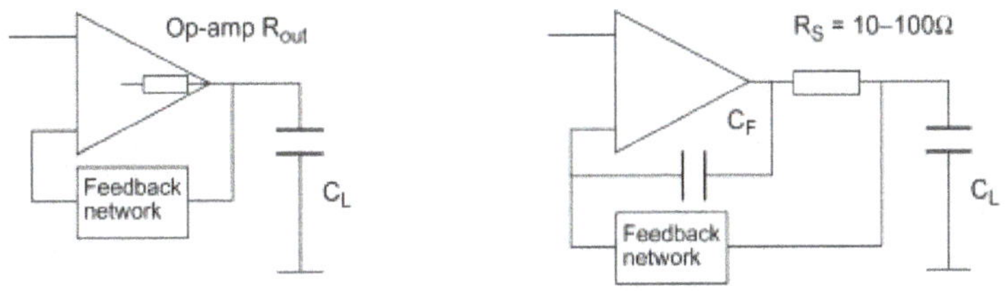

Phase lag @ freq f = $\tan^{-1}[f/f_c]$ degrees where $f_c = 1 / 2\pi \cdot R_{out} \cdot C_L$

Isolate a large value of C_L with R_S and C_F, typically 20pF

Ohm's Law and Basic Analog Circuit Concepts

Resistance: Resistance can be defined as the characteristic of a medium that opposes flow of current through itself. The unit of resistance is ohms which is represented by the Greek letter Ω (Omega). The power value associated with resistance is quantified as the amount of power that the resistor can dissipate as heat without overheating itself.

The current (I) through the resistor (R) is defined as:

$$I=\frac{V}{R}$$

$$V=I.R$$

$$R=\frac{V}{I}$$

For a 1 Megaohm resistance, the current resulting from the application of 10 Volts would be 10 microamperes.

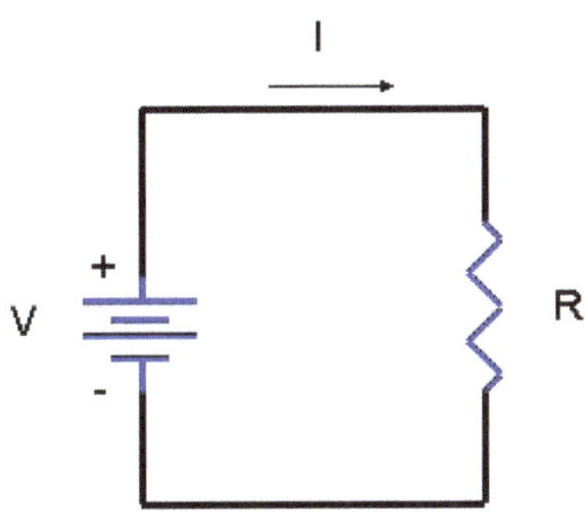

Figure: Simple Representation of Ohm's Law

Ohm's law is the fundamental equation that describes the above relationship between the voltage potential, the current flowing in the circuit, and the resistance of a circuit. The power dissipated in a load resistance (R) is defined as the product of the current and the voltage. Other relationships for power can be easily derived from this by applying Ohm's law using substitution.

The power (P) dissipated in (R) is defined as:

$$P=I.V$$

$$P=\frac{V^2}{R}$$

$$P=I^2.R$$

To calculate the value of resistance that will result in 10 Watts with 10 Volts applied to it, we remember that $P = V^2/R$. Transposing, $R = V^2/P$. The resistance is 100/10, or 10 ohms. So 10V applied to 10 ohms will yield 10 Watts. Any time two of the parameters (V, R, or P) are numerically the same, the third one will be the same. A common way to measure resistance is by using a Digital Multimeter (DMM).

The power dissipation (P) is what sets the limit for how much voltage can be applied to the 50 ohm input of a digitizer. From the equations, we see that 10 Volts into 50 ohms will require the digitizer's input load to dissipate 2 Watts. If you are running 2 channels, that's 4 Watts. This amount of power dissipation is definitely something that can't be ignored. Also notice that because of the square law effect, if you double that voltage into the digitizer, the power it must dissipate will QUADRUPLE.

Voltage Divider Calculation

When two resistors are connected in a series configuration, they must share the applied voltage and the same current flows through both of them.

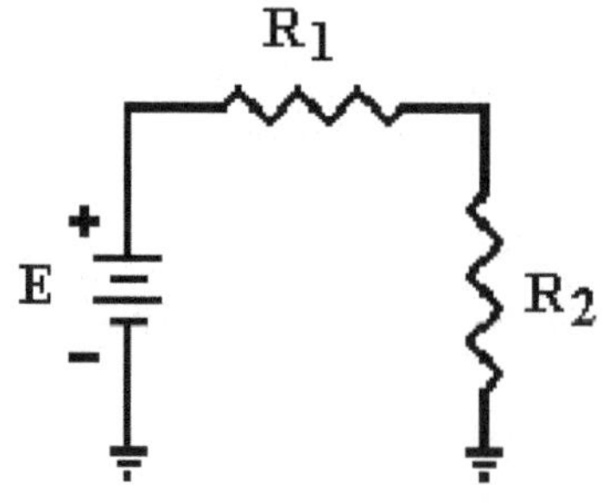

Figure: Voltage Divider Circuit Example

The formula used to calculate the applied voltage is:

$$E_1 = R_1.I$$

(E_1 = Voltage drop across R_1)

$$E_2 = R_2.I$$

(E2 = Voltage drop across R2)

$$R_{eq} = R_1 + R_2$$

$$I = \frac{E}{R_{eq}}$$

$$E = E_1 + E_2$$

$$E = I.(R_1 + R_2)$$

To calculate the voltage across R2:

$$E_2 = R_2.I$$

$$E_2 = R_2.\frac{E}{R_{eq}}$$

$$E_2 = R_2.\frac{E}{(R_1 + R_2)}$$

$$E_2 = E.\frac{R_2}{(R_1 + R_2)}$$

Voltage divider is described by the equation above.

Current Divider Calculation

When two resistors are connected in parallel configuration, the same voltage is across each of them. The amount of current flowing through them depends on the value of the resistances.

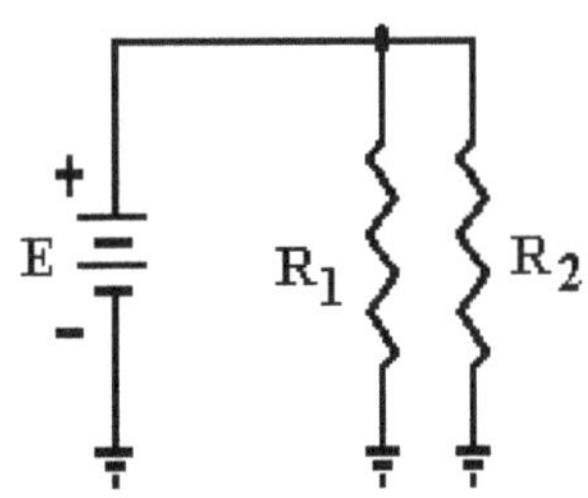

Figure: Current Divider Circuit Example

The above figure depicts two resistors in a parallel configuration.

$$I = I_1 + I_2$$

$$E = I_1.R_1$$

$$E = I_2.R_2$$

$$I = \frac{E}{R_1} + \frac{E}{R_2}$$

$$I = E\left(\frac{1}{R_1} + \frac{1}{R_2}\right)$$

$$E = I.\text{R}_{eq}$$

$$\text{R}_{eq} = \frac{1}{(1/R_1 + 1/R_2)} = \frac{R_1.R_2}{R_1 + R_2}$$

Digital multimeters (DMMs) are the most common measurement devices found in automated test systems. DMMs are usually simple to use and are often low-cost instruments. Generally, DMMs have built-in conditioning that provides:

a) High resolution (commonly measured in digits),

b) Multiple measurements (volts, current, resistance, etc),

c) Isolation and high voltage capabilities.

Capacitance Calculations

Capacitors store energy in the form of electrical charge. Amount of charge that the capacitor can hold depends on the area of the two plates in the figure above and the distance between them. Large plates with a small distance between them have a higher capacity to hold charge. The electric field between the plates of a capacitor resists changes in applied voltage. Capacitors decrease their resistance with frequency.

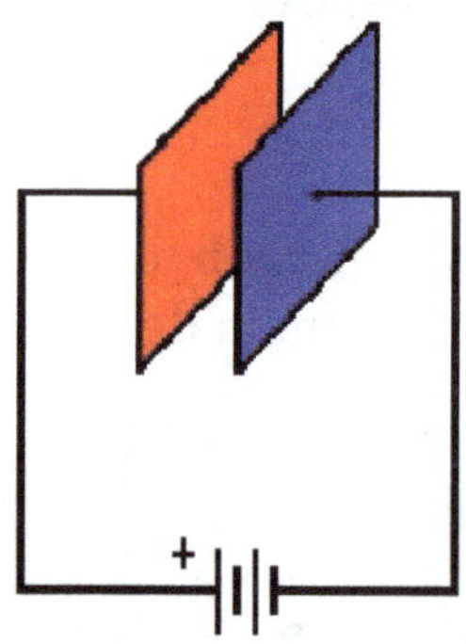

C

Figure: Capacitance Circuit Example

Reading Capacitor Values

The unit of capacitance is Farad which is represented by the letter F. The formula to calculate capacitance is:

$$\text{C} = \frac{\text{Q}}{\text{V}}$$

Where,

C = Capacitance in farads

Q = Accumulated Charge in Coulombs

V = Voltage difference between the plates

Figure: Series Capacitor Configuration

The above configuration represents two capacitors in series.

Since the capacitance of a capacitor is inversely proportional to the distance between the plates, the total capacitance CT of any number of capacitances can be calculated by the following:

$$\left(\frac{1}{C_{rotal}}\right)=\left(\frac{1}{C_1}\right)+\left(\frac{1}{C_2}\right)+\left(\frac{1}{C_3}\right)+\ldots$$

For two capacitors connected in series,

$$C_1=\frac{Q}{V_1}$$

$$C_2=\frac{Q}{V_2}$$

$$V_1=\frac{Q}{C_1}$$

$$V_2=\frac{Q}{C_2}$$

$$V=V_1+V_2$$

$$V=\frac{Q}{C_{eq}}$$

$$C_{eq}=\frac{1}{(1/C_1+1/C_2)}=\frac{C_1.C_2}{C_1+C_2}$$

Parallel Configuration

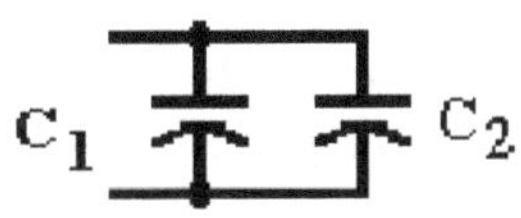

Figure: Parallel Capacitor Configuration

Each capacitor charges to the same applied voltage. Total capacitance equals the sum of the individual capacitances of the capacitors.

The formula used to calculate the capacitance is below:

$$Q_1 = C_1.V$$
$$Q_2 = C_2.V$$
$$Q = Q_1 + Q_2$$
$$Q = V.(C_1 + C_2)$$
$$C_{eq} = C_1 + C_2$$

Inductance Calculations

Inductance is defined as the amount of voltage dropped across the inductor for a given rate of change of current flowing through it. Inductors increase their resistance with frequency. The unit of inductance is Henry which is represented by the letter H.

Series Configuration

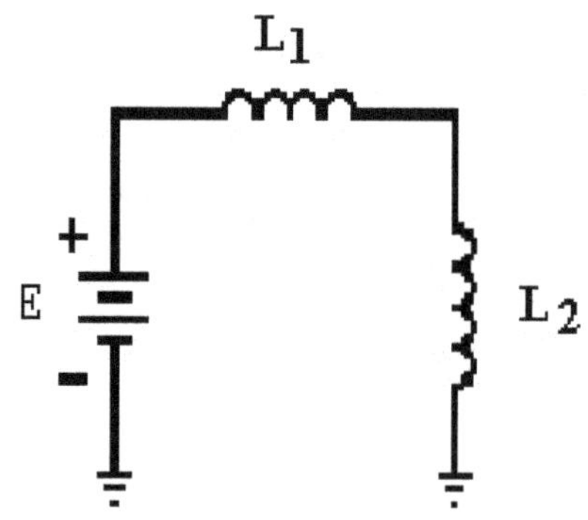

Figure: Series Inductor Configuration

The Figure above is a series configuration of two inductors.When two inductors are connected in series as shown above then their total inductance equals the sum of individual inductances.

$$L_1 = E.\frac{dI}{dt}$$
$$L_2 = E.\frac{dI}{dt}$$
$$L_T = L_1 + L_2$$

Where dI/dt is the change of current over time.

However, in the real world if we consider mutual inductance where magnetic field of each inductor affects the other coil then the total inductance can be calculated using the formula below:

$$L_T = L_1 + L_2 \pm 2.M$$

Where, M is the mutual inductance between the two coils.

Parallel Configuration

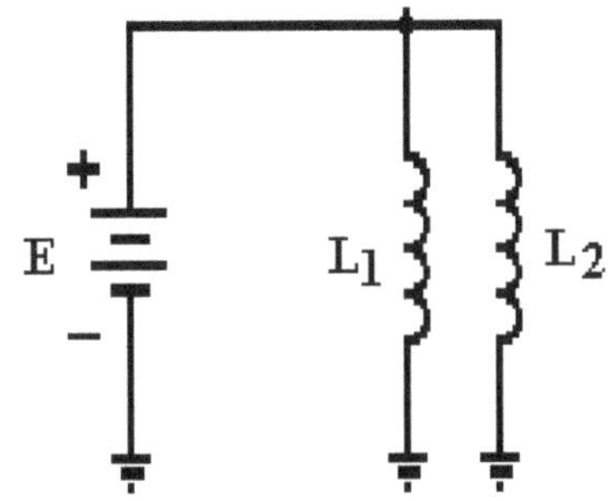

Figure: Parallel Inductor Configuration

Figure above is a parallel configuration of two inductors. When two inductors are connected in parallel configuration, mutual inductance needs to be considered. Also, mutual inductance will be either added or subtracted from the self-inductance of each coil since current has two paths to flow in.

The total inductance can be calculated using the formula below:

$$L_T = \frac{L_1.L_2 - M^2}{L_1 + L_2 - 2.M}$$

The NI PXI-4072 FlexDMM and LCR Meter brings digital multimeters one step closer to providing universal measurement capability by offering engineers the 20 most common automated test measurements, including voltage, current, capacitance, inductance, temperature and resistance.

Impedance

Impedance (Z) is generally defined as the total opposition a device or circuit offers to the flow of an alternating current AC at a given frequency. Its value is equal to the ratio between the voltage and the current over an element of a circuit. Therefore, the unit of impedance is ohm (Ω).

Impedance is represented as a complex quantity which is graphically shown on a vector plane. An impedance vector consists of a real part (resistance, R) and an imaginary part (reactance, X). Impedance can be expressed using the rectangular coordinate form R + jX or in the polar form as a magnitude and phase angle: Z.

Admittance

Admittance (Y) is the reciprocal of impedance. It is also a complex quantity: real part is called conductance (C) and the imaginary part is called susceptance (B).

The unit of admittance is siemens (S):

$$Y = G + jB$$

Where, Y is admittance; G represents conductance; and B represents susceptance.

Analog Amplifier Circuits

Figure below is a basic op-amp model that consists of three basic stages of an op-amp:

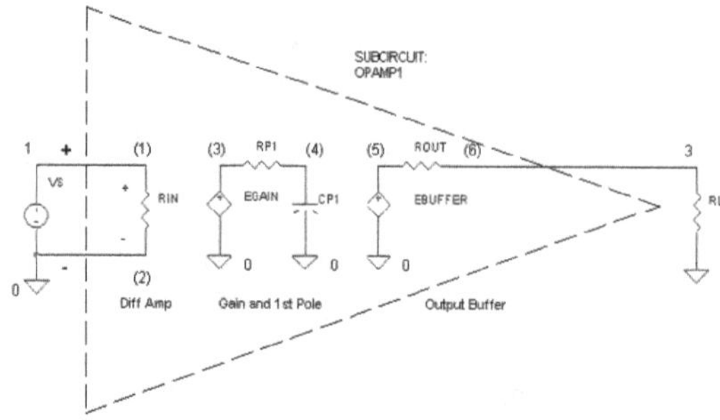

Figure: Basic Operational Amplifier (Op-Amp) Model

1) Differential Amplifier: An amplifier whose output is proportional to the difference between the input signals.

2) Gain/Frequency Response: A filter changes the amplitude or phase characteristics of a signal with respect to frequency. The frequency domain behavior of a filter is mathematically described in terms of a transfer function or a network function. The transfer function H(s) is described as a ratio between output and input signals.

$$H(s) = \frac{V_{out}(s)}{V_{in}(s)}$$

Where, Vout(s) and Vin(s) are the output and input voltage signals and s is the complex frequency variable.

The magnitude of transfer function is called amplitude response or frequency response especially in radio applications.

3) Inverting Amplifier: An Inverting is as simple as it sounds, it simply reverses the polarity of the input signal. For example, if the voltage going into the amplifier is positive, it is negative when it comes out.

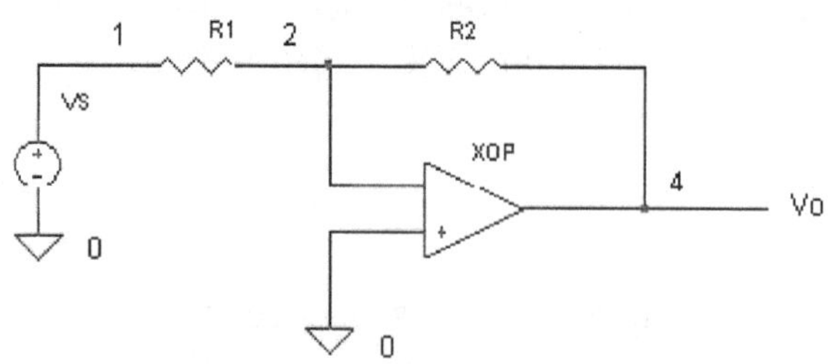

Figure: Basic Inverting Amplifier

Calculating the gain of an inverting amplifier:

$$\frac{V_s - V_1}{R_1} = \frac{V - V_o}{R_2}$$

Since, V1 = V = 0 (Virtual ground)

$$\frac{V_s}{R_1} = \frac{-V_o}{R_2}$$

$$Gain = \frac{V_o}{V_s} = \frac{-R_2}{R_1}$$

Non-Inverting Amplifier: The gain of the amplifer is determined by the ratio of R1 and R2.

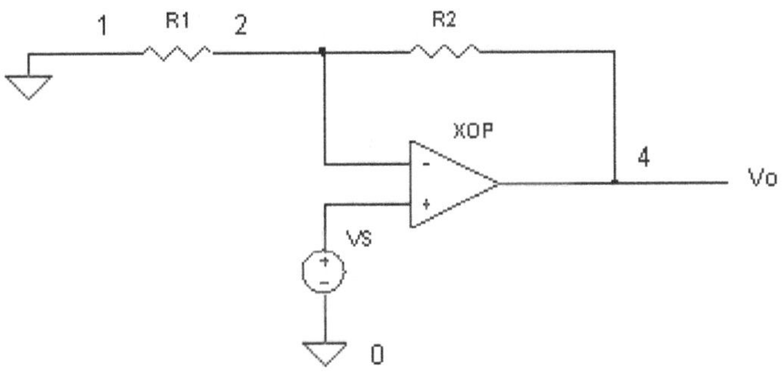

Figure: Basic Non-Inverting Amplifier

Calculating the gain of a non-inverting amplifier:

$$V_o.R_1 = V_s.R_1 + V_s.R_2$$

$$R_1.(V_o - V_s) = V_s.R_2$$

$$\frac{V_o}{V_s} - \frac{R_2}{R_1}$$

$$Gain - 1 = \frac{R_2}{R_1}$$

$$Gain = 1 + \frac{R_2}{R_1}$$

Where,

$$Gain = \frac{V_o}{V_s}$$

All the Data Acquisition (DAQ) products offered by National Instruments have built-in amplifiers.

Analog RC Filters

- RC Low Pass Filter: A common circuit to attenuate high-frequency components in an analog signal is the RC Low Pass Filter. Examine the diagram below, where Vin is the applied voltage and the voltage Vout across C1 is the output.

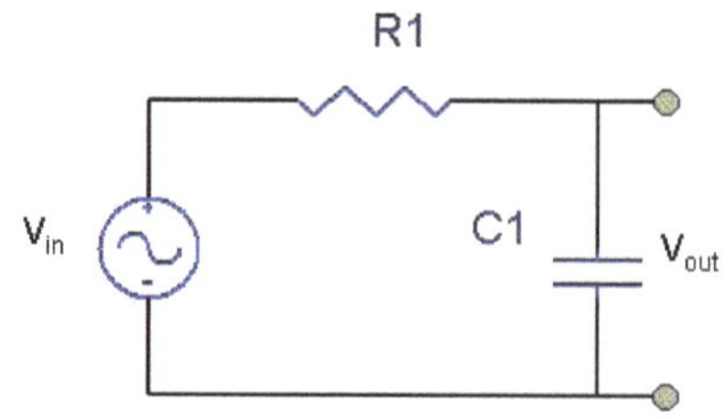

Figure: Simple RC Low Pass Filter

The RC Low Pass Filter passes low frequency and DC signals to the output, but blocks out high frequency signals. This could be either desirable or undesirable.

You get the equivalent of this circuit even when you do not expect it. For example, C could be the input capacitance of a Digital Multimeter (DMM), Digitizer, etc. R in this example might be the source resistance of the device under test (DUT). The DUT has to charge and discharge C as the signal varies. As the variation increases in frequency, the impedance of C eventually becomes lower than R and starts to attenuate the signal. The frequency where the value of Vout is at 0.707 of Vin, is defined as the −3dB frequency or the half-power point, because the output is down −3dB of the input signal at that point.

$$t = R.\mathrm{C}$$

$$\mathrm{F}_{\mathrm{abd}} = \frac{1}{2.\pi.t}$$

- RC High Pass Filter: A circuit that attenuates low-frequency components in an analog signal is called a RC High Pass Filter. Notice that the circuit is similar to the one above, but Vout is now measured across R1.

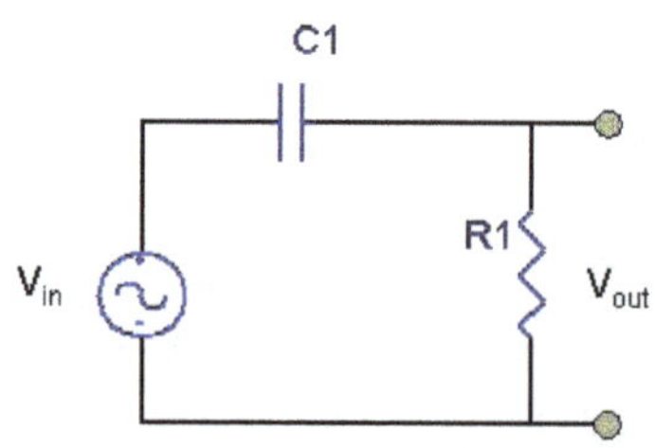

Figure: Simple RC High Pass Filter

A very important typical application of this circuit is in the input coupling circuits of a Digitizer or Digital Multimeter (DMM). With the capacitor in, it is “AC Coupled.” With the capacitor shorted, it would be “DC Coupled.”

Digital Circuit

A Digital Circuit is also a type of an electronic circuit that is predominantly built using Digital electronic components to process digital signals. At low level, the digital circuits consist of a combination of transistors, logic gates (AND, NAND, NOT etc.) and at high level, microcontrollers and processors.

All Digital circuits work on the binary digital signalling i.e. using only two voltage levels to represent the status signal i.e. ON or OFF. For performing various digital signal processing, all digital circuits incorporate fast switching devices like diodes and transistors and some of the important functions of digital circuits are multiplexing, demultiplexing, memory, combinational logic, sequential logic, etc.

The following is an example of a simple digital circuit built using CD4017 Counter IC.

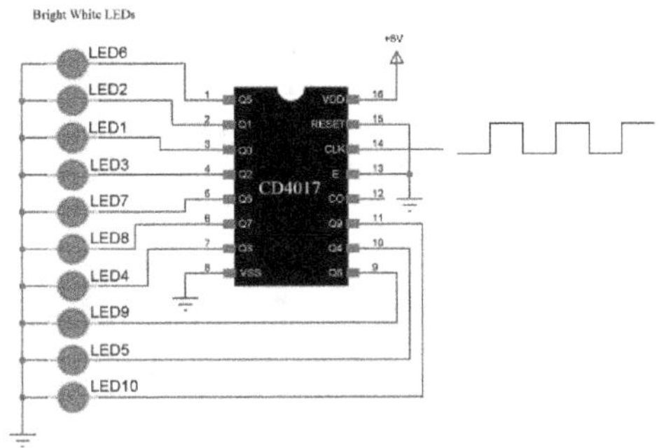

This might not be clear but if you take a look at the internal structure of the CD4017 IC, you will see that it is essentially constructed using AND Gates, NOR Gates, Inverters, Flip-flops and buffers.

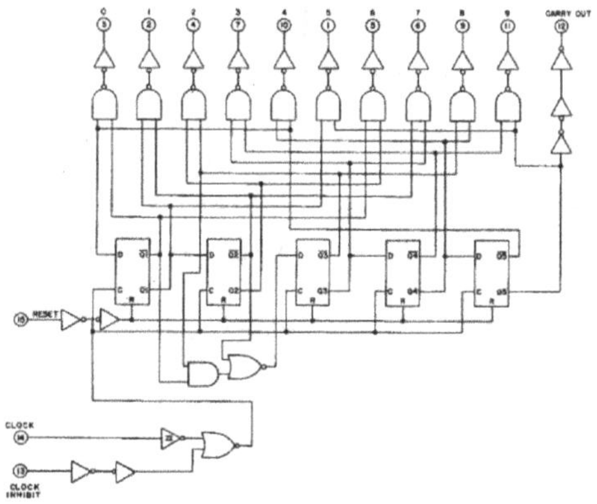

All the modern computers work in digital realm even though they are built using analog electronic components. Since most of the physical world is analog, the analog signals must be converted to digital signals before the computers can perform any operations.

Differences between Analog Circuits and Digital Circuits

At this point, the differences between Analog Circuits and Digital Circuits might be clear but as an aid for easy understanding, we have tabulated the differences.

Analog Circuits	Digital Circuits
Analog circuits operate on continuously variable signals also known as Analog Signals.	Digital Circuits operate on discretely variable signals or Digital signals i.e. the signal exists only in two levels: 0 and 1 (binary digital signalling).
Depending the efficiency and precision, it is quite difficult to design Analog Circuits.	Digital Circuits are relatively easy to design with many automated tools available for various stages of design and analysis.

When interacting with the physical world, analog circuits can directly accept the signals from outside as the data is already analog.	If a digital circuit has to acquire data from physical world, the analog signals must be converted to digital signals first.
As there is no need for data conversion, there is ideally no loss of information.	During the process of converting analog signals to digital signals, there might a significant amount of data loss, which can result in loss of information.
If precision and accuracy are not a criterion, then analog circuits can be simple and inexpensive.	Even with simple design techniques and at low cost, the digital circuits can provide good accuracy and precision.
Due to the lack of skilled engineers and the complexity of the designs, analog circuits can turnout to be quite expensive.	Advanced Integrated Circuits technologies and many other factors help the digital circuits to be reliable, lower in cost and smaller in size.

Digital Logic Circuits

The digital logic circuits are basic building blocks of the digital systems (digital computers). These digital logic circuits can be classified into two categories such as combinational logic circuits and sequential logic circuits.

Combinational Logic Circuits

A combinational logic circuit is one in which the present state of the combination of the logic inputs decides the output. The term combination logic means combining of two or more logic gates to form a required function where the output at a given time depends only on the input.

The logic gates are the fundamental building blocks of a combinational circuit. By using the combination of logic gates more complex combinational circuits can be implemented such as multiplexers and de-multiplexers, comparators, adders and subtractors, etc.

A combinational circuit comprises of input variables, logic gates and output variables. The logic gates accepts the inputs and depending on the type of functioning of the logic gate,output signals are generated from them.

The required output data is obtained from this process by transforming the binary information given at the input. The figure below shows the schematic representation of a generalized combinational logic circuit consisting of n input variable and m output variables.

In the above figure, there are n input variables and hence there will be 2n possible combinations of bits at the input. By a Boolean expression of input variables, each output is expressed. So the result of above generalized combinational logic circuit can be expressed by m Boolean expressions.

In the above figure, the circuit accepts the binary variables and depending on the logical combination of gates it generate outputs.

Design Procedure of Combinational Logic Circuits

A combinational circuit can be designed using the following steps:

1. Identification and determination of number of available input variables and required output variables.
2. Representing symbols (alphabets) for each and every input and output variables.
3. Expressing the input and output variable's relationship.
4. Construction of truth table indicating the relationship between the input and output variables.
5. Obtaining the Boolean expression for each output variable in terms of input variables.
6. Minimizing the Boolean expressions of various output variables.
7. Obtaining the logic diagram by the implementation of minimized Boolean expressions.

For minimizing Boolean expressions, various simplification techniques are available so that they reduce the number of gates, hence less implementation cost. These techniques include Boolean algebra theorems and identities, Karnaugh maps (K-maps), Quinne-McCluskey tabulation, etc. For hardware implementation of the combinational circuit, the following guidelines are mostly preferred.

- The implementation of the circuit should be in such that, it should have minimum number of gates, having a minimum number of inputs.
- The number of interconnections between the gates should be minimum and the propagation time should be shortest.
- There should be always a limitation on the driving capability of the gates.

This is the simple, effective and traditional method of designing a combinational circuit for small circuit. If the circuit is more complex, number of gates required is more and also it requires more number of wires in between them. Hence, designing such circuits may be less reliable and time consuming.

To overcome these problems, most of the combinational circuits are available in integrated circuits (ICs) which are extensively used in the design of the digital systems. Depends on the integration capability of gates, these ICs are classified into small, medium, large and very large scale ICs.

For performing specific digital functions such as addition, multiplexing, de-multiplexing, encoding, decoding, comparison, etc., mostly medium scale integrated (MSI) ICs are used in digital systems.

Classification of Combinational Logic Circuits

Combinational circuits are used in a wide variety applications including calculators, digital measuring techniques, computers, digital processing, automatic control of machines, industrial processing, digital communications, etc.

Different types of combinational logic circuits are used for various applications.Depending on the function of the logic circuit employed for, combinational logic circuits are mainly classified into three types, namely Arithmetic and logic circuits, data transmission circuits and code converter circuits.

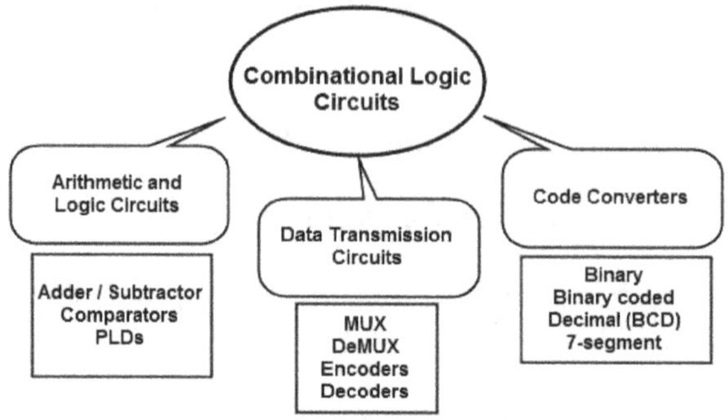

Arithmetic and Logic Circuits

Arithmetic operations are one of the main functionalities of most computers and calculators. These operations are carried by the logic gates or simply combinational circuits which combines the several logic gates to perform the required function. These arithmetic functionalities of the combinational circuits include addition, subtraction, multiplication, etc.

Some of the combinational circuits used for these operations are Half- adder, Full-adder, Half-subtractors, Full subtractors, Adder-subtractors, Comparators, PLDs (Programmable Logic Devices), etc.

Data Transmission Circuits

The most commonly used combinational circuits are multiplexers and de-multiplexers. A multiplexer logic circuit accepts the several data inputs and passes one of them through the output at a time. These are used in data selection, parallel to series conversion, data routing applications in digital systems.

A de-multiplexer circuit performs the reverse operation than that of the multiplexer. It accepts the single input and distributes it to multiple outputs. These are used in distributors as well as series to parallel converter applications.

Other essential data transmission circuits include encoders and decoders. A decoder logic circuit converts n-bit binary input code into 2n output lines. Each output line is activated for only one of the possible combinations of the inputs.

These are used in data de-multiplexing, digital to analog converters and digital display applications. An encoder digital circuit converts an active input signal into a coded output signal as reverse operation of the decoder. These are used in bit compression applications.

Code Converter Circuits

In some applications, it is required to interface two digital blocks of different coding systems. So a conversion circuit is used in between such circuits to convert the information. Some of these converters are binary to gray code, gray to binary, BCD to excess-3 code, Excess-3 to BCD code and seven segment code converter circuits.

Binary Adder

Binary adder is one of the basic combinational logic circuits. The outputs of a combinational logic circuit depend on the present input only. In other words, outputs of combinational logic circuit do not depend upon any previously applied inputs. It does not require any memory like component. Binary adder is one of the basic combinational logic circuits as present state of input variables.

Designing Binary Adder

Half Adder

Before designing a binary adder, let us know some basic rules of binary addition. The most basic binary addition is addition of two single bit binary numbers that is addition of two binary digits.

The binary digits are 0 and 1. Hence, there must be four possible combinations of binary addition of two binary bits:

$$0+0=0$$

$$0+1=1$$

$$1+0=1$$

$$1+1=10$$

In the above list, first three binary operations result in one bit but fourth one result in two bits. In one bit binary addition, if augend and addend are 1, the sum will have two digits. The higher significant bit (HSB) or Left side bit is called carry and the list significant bit (LSB) or right side bit of the result is called sum bit. The logical circuit performs this one bit binary addition is called half adder.

Design of Half Adder

For designing a half adder logic circuit, we first have to draw the truth table for two input variables i.e. the augend and addend bits, two outputs variables carry and sum bits.

In first three binary additions, there is no carry hence the carry in these cases are considered as 0.

Truth Table for Half Adder

Inputs		Outputs	
Augend (A)	Addend (B)	Carry (C)	Sum (S)
0	0	0	0
0	1	0	1
1	0	0	1
1	1	1	0

K-map for Half Adder

Now from this truth table we can draw K-map for carries and sums separately.

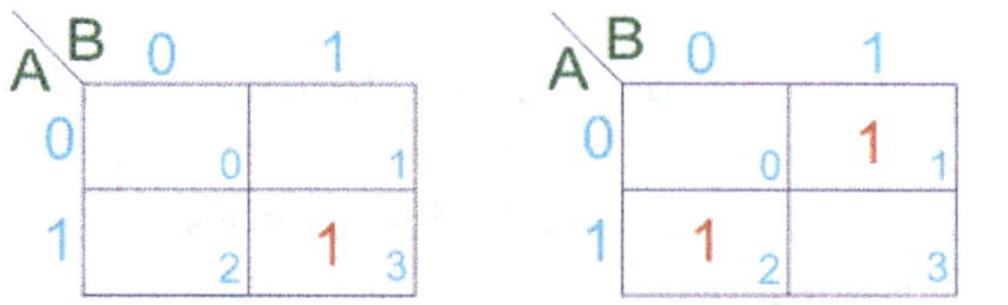

K-map for Carry **K-map for Sum**

For above K-maps we get,

$$Carry\, C = AB \; and \; Sum\, C = \overline{AB} + \overline{AB}$$

Hence, the logical design of Half Adder would be:

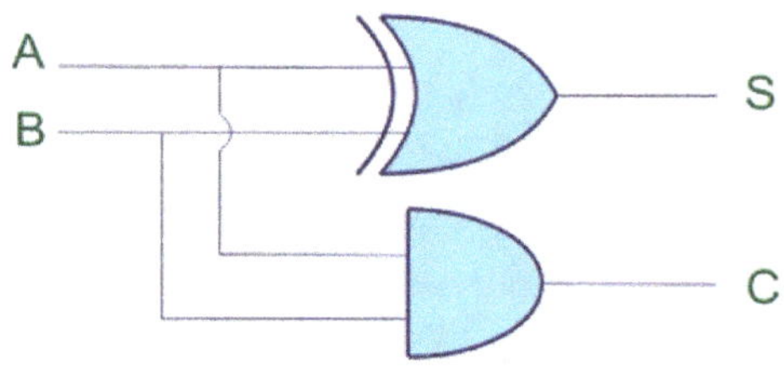

Although from truth table it is clearly seen that carry (C) column signifies AND operation and sum (S) column signifies XOR operation between input variables but till we went through K-map as it is general practice to do so for more complex binary logic operations.

Full Addition

Before knowing about full adder, let us know what is full addition? For that let us consider the example,

$$\begin{array}{r} 1101 \\ {}^{+}0111 \\ \hline 10100 \end{array}$$

There, are two four bits binary numbers 1101 and 0111 which we have to add. The process of binary addition is like follows:

1. We have to add first list significant bit (LSB) of both 4bits binary number first and this will result a two bits binary number.

$$\begin{array}{r} 110\boxed{1} \\ 011\boxed{1} \\ \hline 10 \end{array}$$

Here, LSB of 1101 and 0111 are 1, hence 1 + 1 = 10. The LSB of 10 is 0 and higher significant bit (HSB) is 1.

2. The LSB of the result is sum and to be put at the list significant position of the final result of the sum, and HSB of the two bits results will be carry and to be added with next higher significant bit of two 4bits augend and addend are 0 and 1 and the carry of result i.e. 1 to be added with 0 and 1.

   $0+1+1=10$

3. After this addition, that is next higher than list significant bit of bits of both binary augend and addend and it is carry we get another two bits result. This also has carry and sum. Here also we will write sum at final result and add the carry to the next higher significant bits of augend and addend. This will continue up to most significant bit of augend and addend.

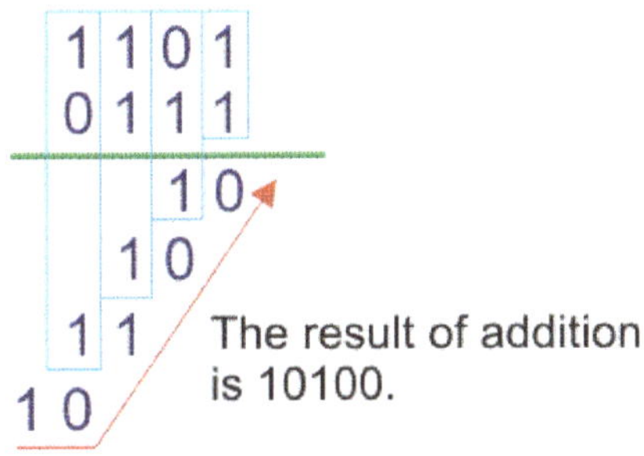

Full Adder

Full adder is a conditional circuit which performs full binary addition that means it adds two bits and a carry and outputs a sum bit and a carry bit.

Any bit of augend can either be 1 or 0 and we can represent with variable A, similarly any bit of addend we represent with variable B. The carry after addition of same significant bit of augend and addend can represent by C. Hence truth table for all combinations of A, B and C is as follows,

SL No	Augend (A)	Addend (B)	Carry (C)	SUM (S)	Final Carry (C_{out})
0	0	0	0	0	0
1	0	0	1	1	0
2	0	1	0	1	0
3	0	1	1	0	1
4	1	0	0	1	0
5	1	0	1	0	1
6	1	1	0	0	1
7	1	1	1	1	1

From the above table, we can draw K-map for sum (s) and final carry (Cout).

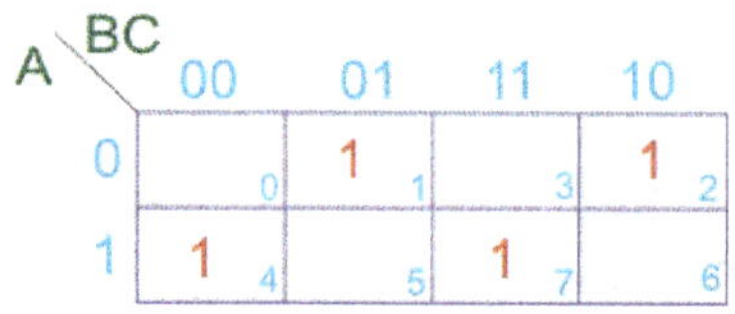

K-map for Sum (S)

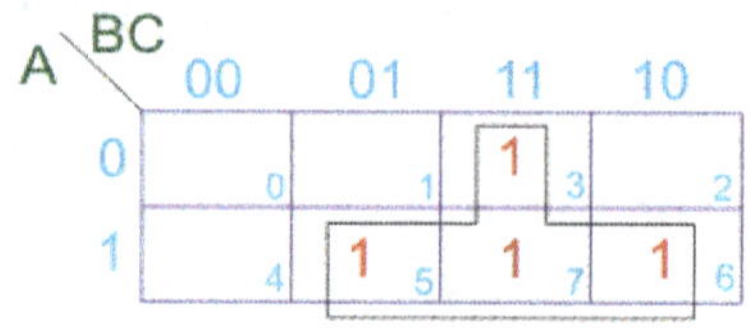

K-map for Carry (C_{out})

Hence, from K-maps,

$$\begin{aligned} S &= A\overline{B}\,\overline{C} + \overline{A}\,\overline{B}C + ABC + \overline{A}B\overline{C} \\ &= C(AB + \overline{A}\,\overline{B}) + \overline{C}(\overline{A}B + A\overline{B}) \\ &= C(\overline{\overline{A}B + A\overline{B}}) + \overline{C}(\overline{A}B + A\overline{B}) \\ &= C(\overline{A \oplus B}) + \overline{C}(A \oplus B) = A \oplus B \oplus C. \end{aligned}$$

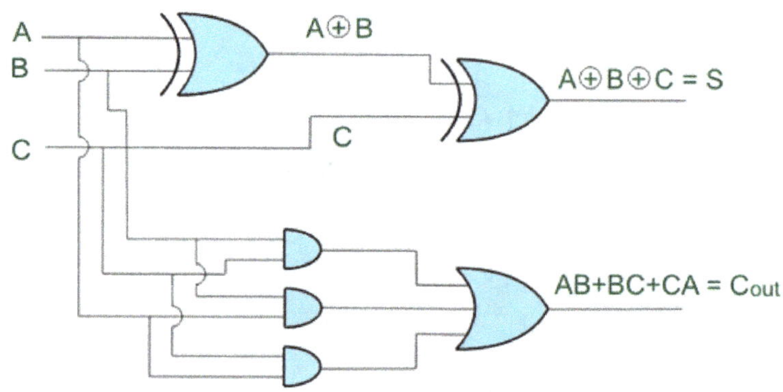

$$\begin{aligned} C_{out} &= \overline{A}BC + A\overline{B}C + AB\overline{C} + ABC \\ &= (\overline{A}B + A\overline{B})C + AB(\overline{C} + C) \\ &= (A \oplus B).C + AB. \end{aligned}$$

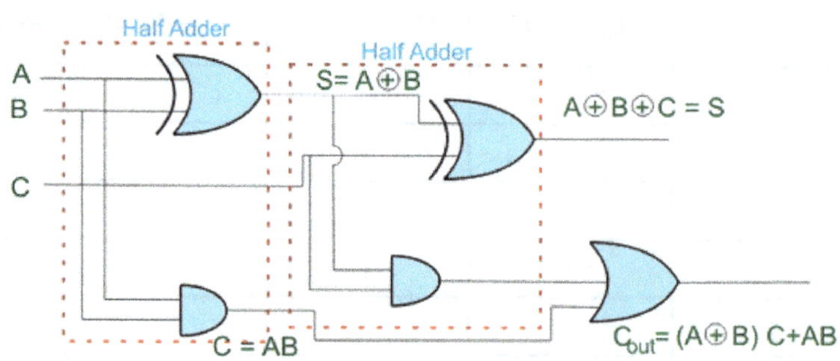

Binary Parallel Adder

A full binary adder performs addition of any single bit of one binary number, same significant or same position bit of another binary numbers and carry comes from result of addition of right side bits of both binary numbers. But a single full adder cannot add more than one bits binary number instantly. This can be done only by connecting as many full adders as the number of bits of the binary numbers whose addition is to be performed. This parallel combination of full adders which performs addition of specific bits binary numbers is called binary parallel adder. For adding two 4 bit binary numbers we have to connect 4 full adders to make 4 bit parallel adder.

The inter connection of 4 full adder in 4bit parallel adder is shown below:

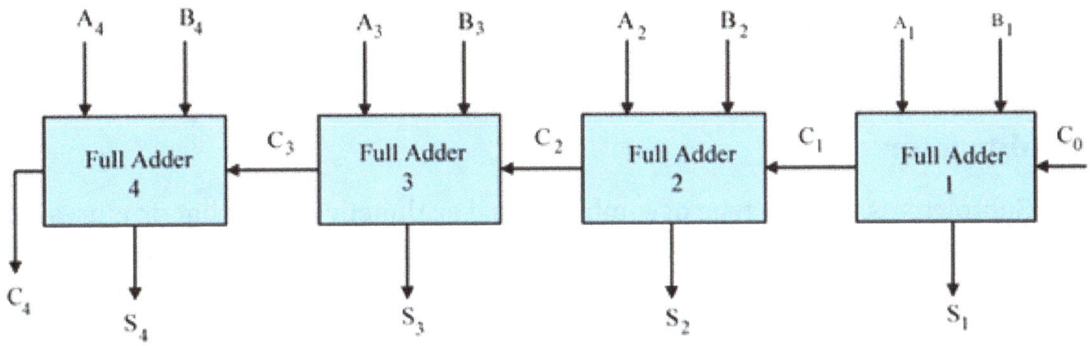

Let us examine the justification of the above circuit by taking an example of addition of two 4 bit binary numbers.

Let us add 1011 with 1101.

$$\begin{array}{r} 1011 \\ 1101 \\ \hline 11000 \end{array}$$

$$\begin{aligned} Here, A_1 &= 1, A_2 = 1, A_3 = 0 A_4 = 1. \\ B_1 &= 1, B_2 = 0, B_3 = 1 B_4 = 1. \end{aligned}$$

As there is no carry $C_0 = 0$.

$$\begin{aligned} Now, C_0 + A_1 + B_1 &= 0+1+1 = 10 \rightarrow S_1 = 0, C_1 = 1 \\ C_1 + A_2 + B_2 &= 1+1+0 = 10 \rightarrow S_2 = 0, C_2 = 1 \\ C_2 + A_3 + B_3 &= 1+0+1 = 10 \rightarrow S_3 = 0, C_3 = 1 \\ C_3 + A_4 + B_4 &= 1+1+1 = 10 \rightarrow S_4 = 1, C_4 = 1 \end{aligned}$$

Therefore, final result of the addition would be:

$$C_4 S_4 S_3 S_2 S_1 = 11000$$

There are 1 bit, 2 bits and 4 bits parallel adders ICs commercially available in market. For n bit parallel adder required number of such ICs are connected together.

4 bit parallel adder IC is 4008. In n bit parallel adder, output carry terminal of one IC would be connected with input carry terminal of next IC.

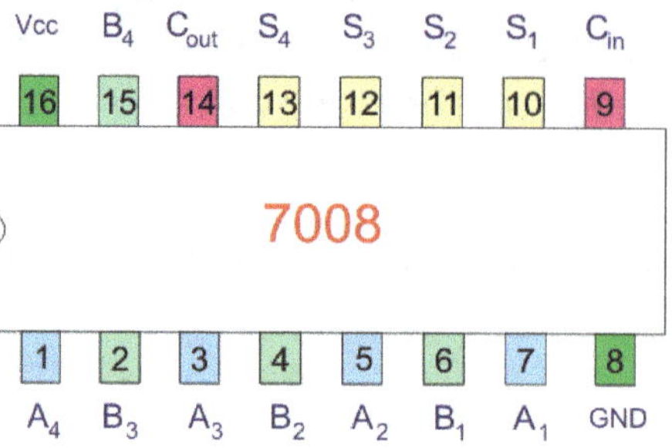

Binary Subtractor

The Binary Subtractor is another type of combinational arithmetic circuit that produces an output which is the subtraction of two binary numbers, Binary Subtractor is a decision making circuit that subtracts two binary numbers from each other, for example, X – Y to find the resulting difference between the two numbers.

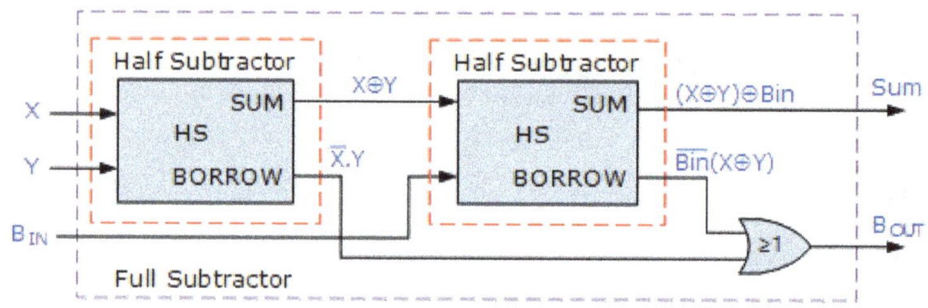

Unlike the Binary Adder which produces a SUM and a CARRY bit when two binary numbers are added together, the binary subtractor produces a DIFFERENCE, D by using a BORROW bit, B from the column. Then obviously, the operation of subtraction is the opposite to that of addition.

the minus sign, "–" is used for a subtraction calculation, and when one number is subtracted from another, a borrow is required if the subtrahend is greater than the minuend. Consider the simple subtraction of the two denary (base 10) numbers below:

$$\begin{array}{lll} 123 & \text{X} & \text{(Minuend)} \\ \underline{-78} & \text{Y} & \text{(Subtrahend)} \\ 45 & \textit{DIFFERENCE} & \end{array}$$

We can not directly subtract 8 from 3 in the first column as 8 is greater than 3, so we have to borrow a 10, the base number, from the next column and add it to the minuend to produce 13 minus 8. This "borrowed" 10 is then return back to the subtrahend of the next column once the difference is found. borrow a 10 if needed, find the difference and return the borrow.

The subtraction of one binary number from another is exactly the same idea as that for subtracting two decimal numbers but as the binary number system is a Base-2 numbering system which uses "0" and "1" as its two independent digits, large binary numbers which are to be subtracted from each other are therefore represented in terms of "0's" and "1's".

Binary Subtraction

Binary Subtraction can take many forms but the rules for subtraction are the same whichever process you use. As binary notation only has two digits, subtracting a "0" from a "0" or a "1" leaves the result unchanged as 0-0 = 0 and 1-0 = 1. Subtracting a "1" from a "1" results in a "0", but subtracting a "1" from a "0" requires a borrow. In other words 0 – 1 requires a borrow.

Binary Subtraction of Two Bits

0	1	1	(borrow)1 → 0
−0	−0	−1	−1
0	1	0	1

For the simple 1-bit subtraction problem above, if the borrow bit is ignored the result of their binary subtraction resembles that of an Exclusive-OR Gate.

2-input Exclusive-OR Gate

Symbol	Truth Table		
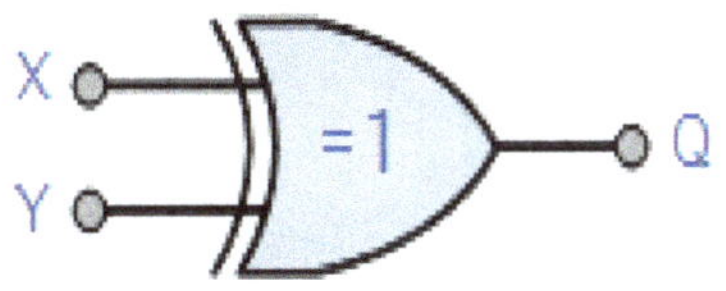	Y	X	Q
	0	0	0
	0	1	1
	1	0	1
2-input Ex-OR Gate	1	1	0

As with the Binary Adder, the difference between the two digits is only a "1" when these two inputs are not equal as given by the Ex-OR expression. However, we need an additional output to produce the borrow bit when input X = 0 and Y = 1. Unfortunately there are no standard logic gates that will produce an output for this particular combination of X and Y inputs.

But we know that an AND Gate produces an output "1" when both of its inputs X and Y are "1" (HIGH) so if we use an inverter or NOT Gate to complement the input X before it is fed to the AND gate, we can produce the required borrow output when X = 0 and Y = 1 as shown below.

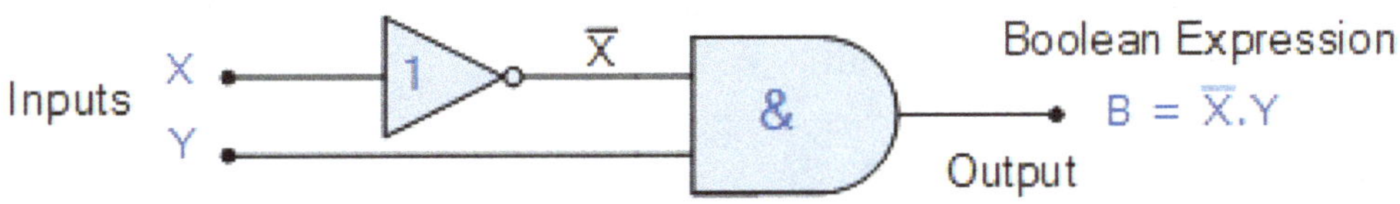

Then by combining the Exclusive-OR gate with the NOT-AND combination results in a simple digital binary subtractor circuit known commonly as the Half Subtractor as shown.

A Half Subtractor Circuit

A half subtractor is a logical circuit that performs a subtraction operation on two binary digits. The half subtractor produces a sum and a borrow bit for the next stage.

Half Subtractor with Borrow-out

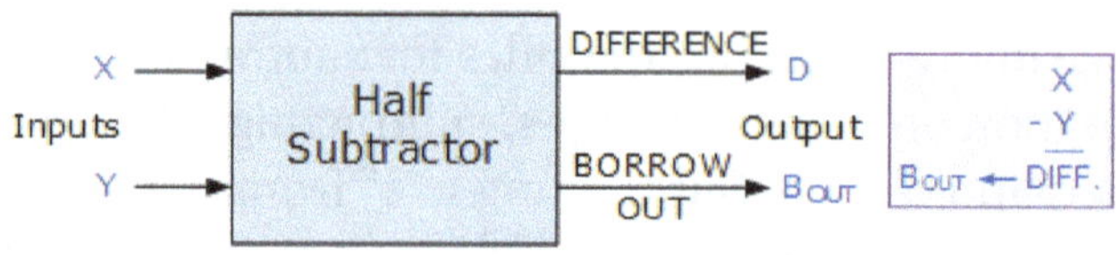

<table>
<tr><th>Symbol</th><th colspan="4">Truth Table</th></tr>
<tr><td rowspan="5">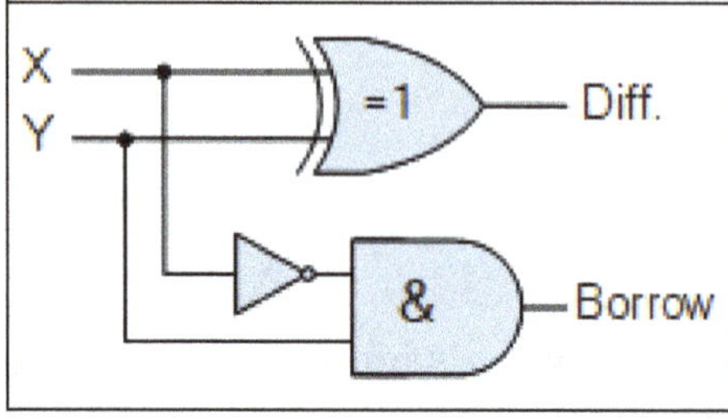</td><td>Y</td><td>X</td><td>DIFFERENCE</td><td>BORROW</td></tr>
<tr><td>0</td><td>0</td><td>0</td><td>0</td></tr>
<tr><td>0</td><td>1</td><td>1</td><td>0</td></tr>
<tr><td>1</td><td>0</td><td>1</td><td>1</td></tr>
<tr><td>1</td><td>1</td><td>0</td><td>0</td></tr>
</table>

From the truth table of the half subtractor the DIFFERENCE (D) output is the result of the Exclusive-OR gate and the Borrow-out (Bout) is the result of the NOT-AND combination. Then the Boolean expression for a half subtractor is as follows:

For the DIFFERENCE bit,

$$D = X \text{ XOR } Y = X \oplus Y$$

For the BORROW bit,

$$B = \text{not-}X \text{ AND } Y = \overline{X}.Y$$

If we compare the Boolean expressions of the half subtractor with a half adder, the two expressions for the SUM (adder) and DIFFERENCE (subtractor) are exactly the same and so they should be because of the Exclusive-OR gate function. The two Boolean expressions for the binary subtractor BORROW is also very similar to that for the adders CARRY. Then all that is needed to convert a half adder to a half subtractor is the inversion of the minuend input X.

One major disadvantage of the Half Subtractor circuit when used as a binary subtractor, is that there is no provision for a "Borrow-in" from the circuit when subtracting multiple data bits from each other. Then we need to produce what is called a "full binary subtractor" circuit to take into account this borrow-in input from a circuit.

A Full Binary Subtractor Circuit

The main difference between the Full Subtractor and the Half Subtractor circuit is that a full subtractor has three inputs. The two single bit data inputs X (minuend) and Y (subtrahend) the same as before plus an additional Borrow-in (B-in) input to receive the borrow generated by the subtraction process from a stage as shown below.

Full Subtractor Block Diagram

Then the combinational circuit of a "full subtractor" performs the operation of subtraction on three binary bits producing outputs for the difference D and borrow B-out. Just like the binary adder circuit, the full subtractor can also be thought of as two half subtractors connected together, with the first half subtractor passing its borrow to the second half subtractor as follows.

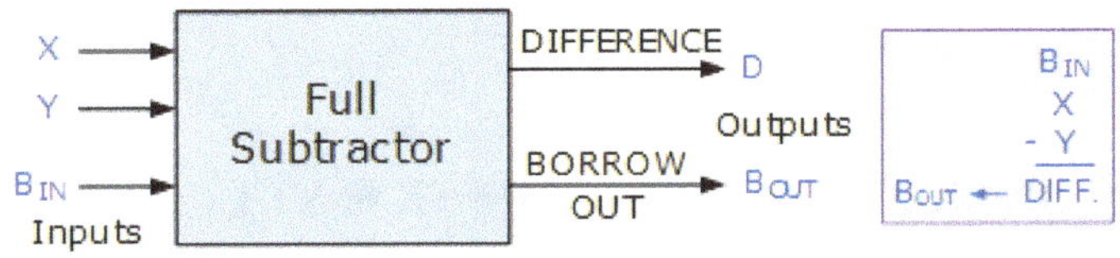

Full Subtractor Logic Diagram

As the full subtractor circuit above represents two half subtractors cascaded together, the truth table for the full subtractor will have eight different input combinations as there are three input variables, the data bits and the Borrow-in, BIN input. Also includes the difference output, D and the Borrow-out, BOUT bit.

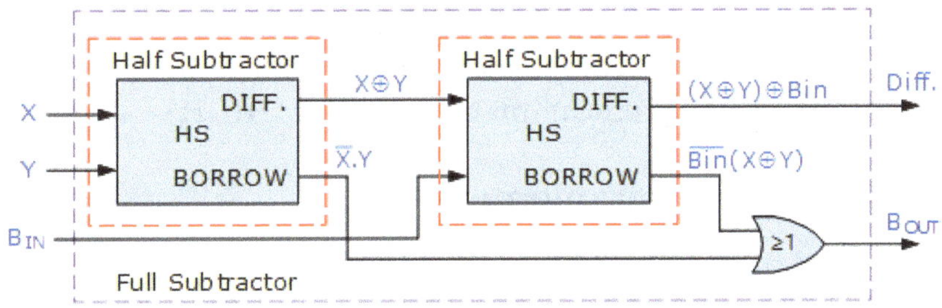

Full Subtractor Truth Table

Symbol	Truth Table				
	B-in	Y	X	Diff.	B-out
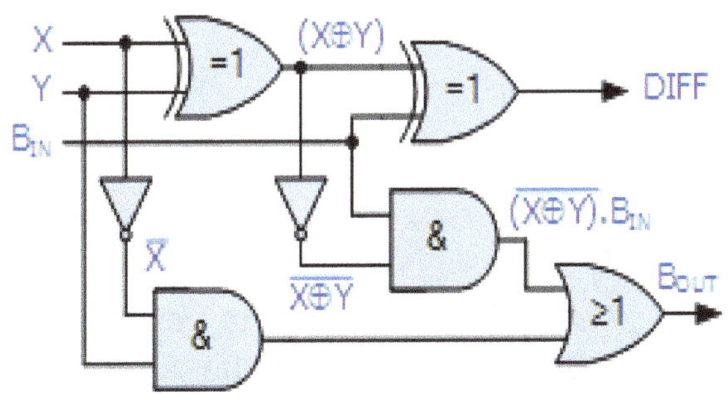	0	0	0	0	0
	0	0	1	1	0
	0	1	0	1	1
	0	1	1	0	0
	1	0	0	1	1
	1	0	1	0	0
	1	1	0	0	1
	1	1	1	1	1

Then the Boolean expression for a full subtractor is as follows:

For the DIFFERENCE (D) bit:

$$D=\left(\overline{X}.\overline{Y}.B_{IN}\right)+\left(\overline{X}.Y.\overline{B_{IN}}\right)+\left(X.\overline{Y}.\overline{B_{IN}}\right)+\left(X.Y.B_{IN}\right)$$

which can be simplified too:

$$D=\left(X\ XOR\ Y\right)\ XOR\ B_{IN}=(X\oplus Y)\oplus B_{IN}$$

For the BORROW OUT (B_{OUT}) bit:

$$B_{OUT} = \left(\overline{X}.\overline{Y}.B_{\overline{IN}}\right) + \left(\overline{X}.Y.\overline{B_{IN}}\right) + \left(\overline{X}.Y.B_{IN}\right) + \left(X.Y.B_{IN}\right)$$

which will also simplify too:

$$B_{OUT} = \overline{X} \text{ AND } Y \text{ OR } \overline{(X \text{ XOR } Y)}B_{IN} = \overline{X}.Y + \overline{(X \oplus Y)}B_{IN}$$

An n-bit Binary Subtractor

As with the binary adder, we can also have n number of 1-bit full binary subtractor connected or "cascaded" together to subtract two parallel n-bit numbers from each other. For example two 4-bit binary numbers. We said before that the only difference between a full adder and a full subtractor was the inversion of one of the inputs.

So by using an n-bit adder and n number of inverters (NOT Gates), the process of subtraction becomes an addition as we can use two's complement notation on all the bits in the subtrahend and setting the carry input of the least significant bit to a logic "1" (HIGH).

Binary Subtractor using 2's Complement

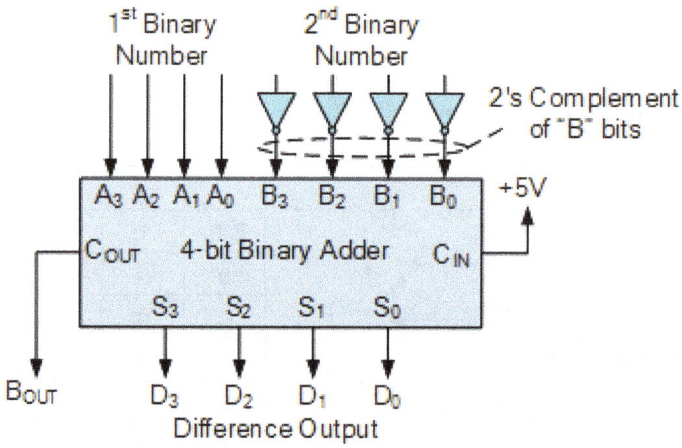

Then we can use a 4-bit full-adder ICs such as the 74LS283 and CD4008 to perform subtraction simply by using two's complement on the subtrahend, B inputs as X – Y is the same as saying, X + (-Y) which equals X plus the two's complement of Y.

If we wanted to use the 4-bit adder for addition once again, all we would need to do is set the carry-in (CIN) input LOW at logic "0". Because we can use the 4-bit adder IC such as the 74LS83 or 74LS283 as a full-adder or a full-subtractor they are available as a single adder/subtractor circuit with a single control input for selecting between the two operations.

Encoder

An Encoder is a combinational circuit that performs the reverse operation of Decoder. It has maximum of 2^n input lines and 'n' output lines. It will produce a binary code equivalent to the input, which is active High. Therefore, the encoder encodes 2^n input lines with 'n' bits. It is optional to represent the enable signal in encoders.

4 to 2 Encoder

Let 4 to 2 Encoder has four inputs Y_3, Y_2, Y_1 & Yo and two outputs A_1 & A_0. The block diagram of 4 to 2 Encoder is shown in the following figure.

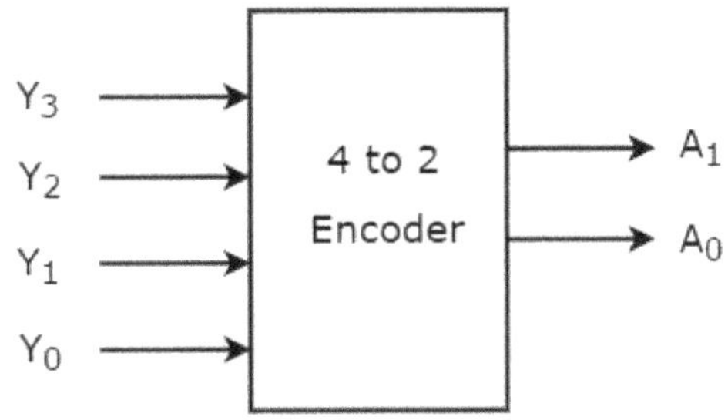

At any time, only one of these 4 inputs can be '1' in order to get the respective binary code at the output. The Truth table of 4 to 2 encoder is shown below:

Inputs				Outputs	
Y_3	Y_2	Y_1	Y_0	A_1	A_0
0	0	0	1	0	0
0	0	1	0	0	1
0	1	0	0	1	0
1	0	0	0	1	1

From Truth table, we can write the Boolean functions for each output as:

$$A_1 = Y_3 + Y_2$$
$$A_0 = Y_3 + Y_1$$

We can implement the above two Boolean functions by using two input OR gates. The circuit diagram of 4 to 2 encoder is shown in the following figure.

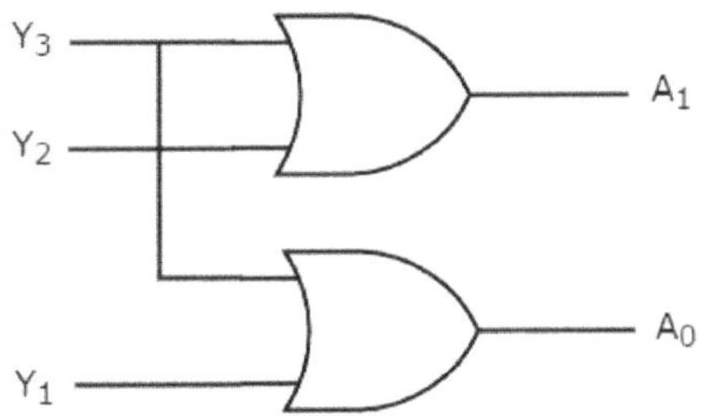

The above circuit diagram contains two OR gates. These OR gates encode the four inputs with two bits.

Octal to Binary Encoder

Octal to binary Encoder has eight inputs, Y_7 to Y_0 and three outputs A_2, A_1 & A_0. Octal to binary encoder is nothing but 8 to 3 encoder. The block diagram of octal to binary Encoder is shown in the following figure.

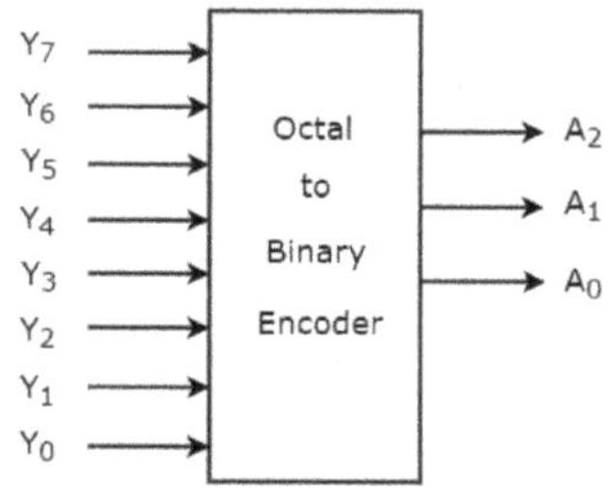

At any time, only one of these eight inputs can be '1' in order to get the respective binary code. The Truth table of octal to binary encoder is shown below in table.

Inputs								Outputs		
Y_7	Y_6	Y_5	Y_4	Y_3	Y_2	Y_1	Y_0	A_2	A_1	A_0
0	0	0	0	0	0	0	1	0	0	0
0	0	0	0	0	0	1	0	0	0	1
0	0	0	0	0	1	0	0	0	1	0
0	0	0	0	1	0	0	0	0	1	1
0	0	0	1	0	0	0	0	1	0	0
0	0	1	0	0	0	0	0	1	0	1
0	1	0	0	0	0	0	0	1	1	0
1	0	0	0	0	0	0	0	1	1	1

From Truth table, we can write the Boolean functions for each output as:

$$A_2 = Y_7 + Y_6 + Y_5 + Y_4$$
$$A_1 = Y_7 + Y_6 + Y_3 + Y_2$$
$$A_0 = Y_7 + Y_5 + Y_3 + Y_1$$

We can implement the above Boolean functions by using four input OR gates. The circuit diagram of octal to binary encoder is shown in the following figure.

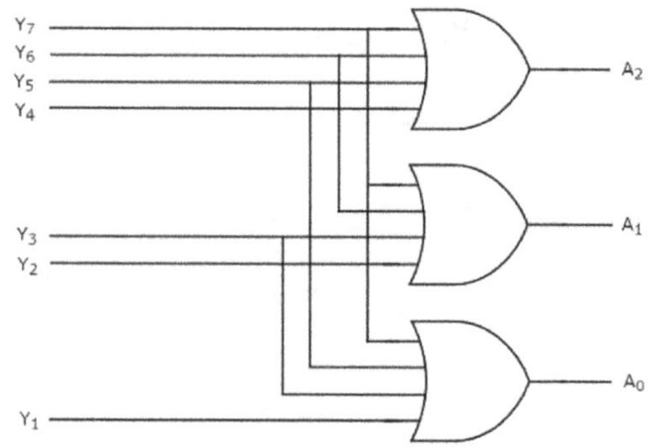

The above circuit diagram contains three 4-input OR gates. These OR gates encode the eight inputs with three bits.

Drawbacks of Encoder

Following are the drawbacks of normal encoder:

- There is an ambiguity, when all outputs of encoder are equal to zero. Because, it could be the code corresponding to the inputs, when only least significant input is one or when all inputs are zero.
- If more than one input is active High, then the encoder produces an output, which may not be the correct code. For example, if both Y_3 and Y_6 are '1', then the encoder produces 111 at the output. This is neither equivalent code corresponding to Y_3, when it is '1' nor the equivalent code corresponding to Y_6, when it is '1'.

So, to overcome these difficulties, we should assign priorities to each input of encoder. Then, the output of encoder will be the (binary) code corresponding to the active High input(s), which has higher priority. This encoder is called as priority encoder.

Priority Encoder

A 4 to 2 priority encoder has four inputs Y_3, Y_2, Y_1 & Y_0 and two outputs A_1 & A_0. Here, the input, Y_3 has the highest priority, whereas the input, Y_0 has the lowest priority. In this case, even if more than one input is '1' at the same time, the output will be the (binary) code corresponding to the input, which is having higher priority.

We considered one more output, V in order to know, whether the code available at outputs is valid or not.

- If at least one input of the encoder is '1', then the code available at outputs is a valid one. In this case, the output, V will be equal to 1.
- If all the inputs of encoder are '0', then the code available at outputs is not a valid one. In this case, the output, V will be equal to 0.

The Truth table of 4 to 2 priority encoder is shown below.

Inputs				Outputs		
Y_3	Y_2	Y_1	Y_0	A_1	A_0	V
0	0	0	0	0	0	0
0	0	0	1	0	0	1
0	0	1	x	0	1	1
0	1	x	x	1	0	1
1	x	x	x	1	1	1

Use 4 variable K-maps for getting simplified expressions for each output.

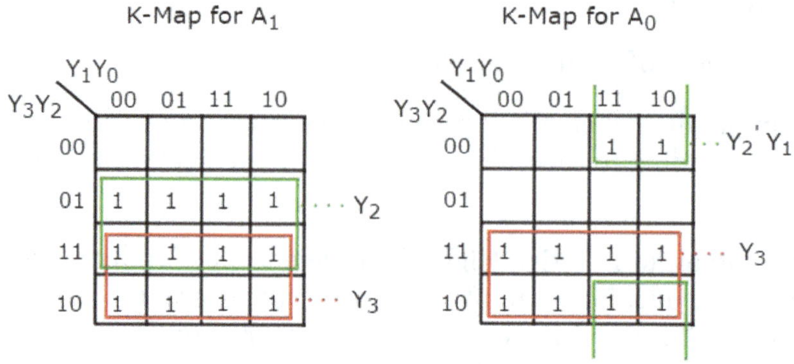

The simplified Boolean functions are,

$$A_1 = Y_3 + Y_2$$

$$A_0 = Y_3 + Y_2'Y_1$$

Similarly, we will get the Boolean function of output, V as,

$$V = Y_3 + Y_2 + Y_1 + Y_0$$

We can implement the above Boolean functions using logic gates. The circuit diagram of 4 to 2 priority encoder is shown in the following figure.

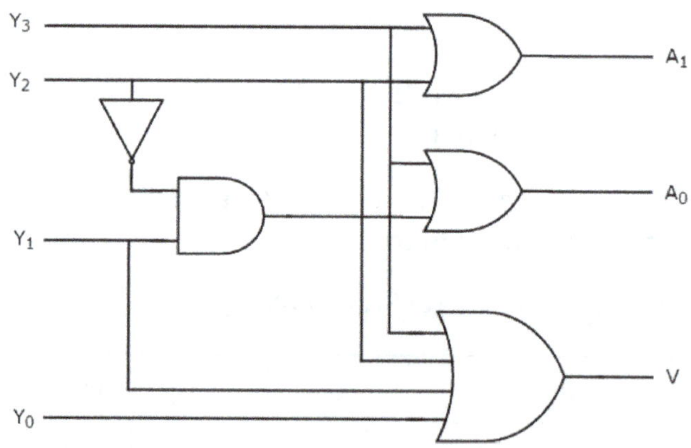

The above circuit diagram contains two 2-input OR gates, one 4-input OR gate, one 2input AND gate & an inverter. Here AND gate & inverter combination are used for producing a valid code at the outputs, even when multiple inputs are equal to '1' at the same time. Hence, this circuit encodes the four inputs with two bits based on the priority assigned to each input.

Multiplexer

The multiplexer is a combinational logic circuit designed to switch one of several input lines to a single common output line, Multiplexing is the generic term used to describe the operation of sending one or more analogue or digital signals over a common transmission line at different times or speeds and as such, the device we use to do just that is called a Multiplexer.

The multiplexer, shortened to "MUX" or "MPX", is a combinational logic circuit designed to switch one of several input lines through to a single common output line by the application of a control signal. Multiplexers operate like very fast acting multiple position rotary switches connecting or controlling multiple input lines called "channels" one at a time to the output.

Multiplexers, or MUX's, can be either digital circuits made from high speed logic gates used to switch digital or binary data or they can be analogue types using transistors, MOSFET's or relays to switch one of the voltage or current inputs through to a single output.

The most basic type of multiplexer device is that of a one-way rotary switch as shown.

Basic Multiplexing Switch

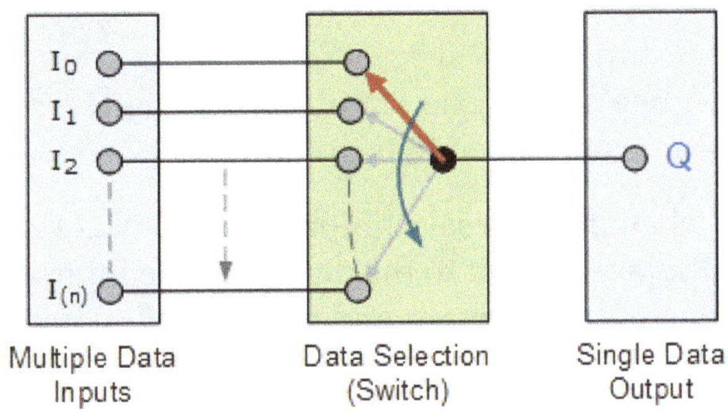

The rotary switch, also called a wafer switch as each layer of the switch is known as a wafer, is a mechanical device whose input is selected by rotating a shaft. In other words, the rotary switch is a manual switch that you can use to select individual data or signal lines simply by turning its inputs "ON" or "OFF". So how can we select each data input automatically using a digital device.

In digital electronics, multiplexers are also known as data selectors because they can "select" each input line, are constructed from individual Analogue Switches encased in a single IC package as opposed to the "mechanical" type selectors such as normal conventional switches and relays.

They are used as one method of reducing the number of logic gates required in a circuit design or when a single data line or data bus is required to carry two or more different digital signals. For example, a single 8-channel multiplexer.

Generally, the selection of each input line in a multiplexer is controlled by an additional set of inputs called control lines and according to the binary condition of these control inputs, either "HIGH" or "LOW" the appropriate data input is connected directly to the output. Normally, a multiplexer has an even number of 2^n data input lines and a number of "control" inputs that correspond with the number of data inputs.

Multiplexers are different in operation to Encoders. Encoders are able to switch an n-bit input pattern to multiple output lines that represent the binary coded (BCD) output equivalent of the active input.

We can build a simple 2-line to 1-line (2-to-1) multiplexer from basic logic NAND gates as shown.

2-input Multiplexer Design

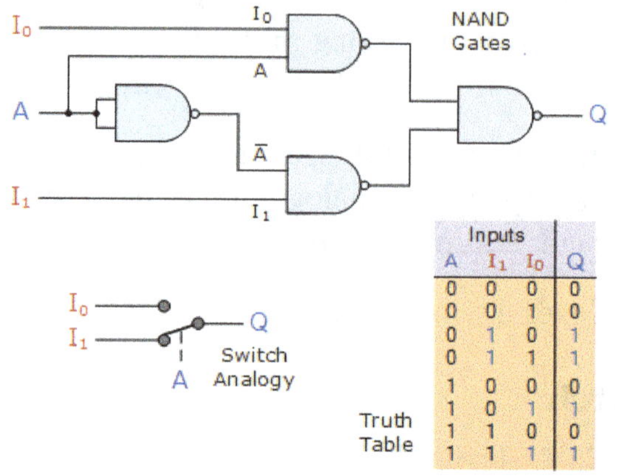

Inputs			
A	I_1	I_0	Q
0	0	0	0
0	0	1	0
0	1	0	1
0	1	1	1
1	0	0	0
1	0	1	1
1	1	0	0
1	1	1	1

The input A of this simple 2-1 line multiplexer circuit constructed from standard NAND gates acts to control which input (I_0 or I_1) gets passed to the output at Q.

From the truth table above, when the data select input, A is LOW at logic 0, input I_1 passes its data through the NAND gate multiplexer circuit to the output, while input I_0 is blocked. When the data select A is HIGH at logic 1, the reverse happens and now input I_0 passes data to the output Q while input I_1 is blocked.

So by the application of either a logic "0" or a logic "1" at A we can select the appropriate input, I_0 or I_1 with the circuit acting a bit like a single pole double throw (SPDT) switch.

As we only have one control line, (A) then we can only switch 2^1 inputs and in this simple example, the 2-input multiplexer connects one of two 1-bit sources to a common output, producing a 2-to-1-line multiplexer. We can confirm this in the following Boolean expression:

$$Q = \overline{A}.\overline{I_0}.I_1 + \overline{A}.I_0.I_1 + A.I_0.\overline{I_1} + A.I_0.I_1$$

and for our 2-input multiplexer circuit above, this can be simplified too:

$$Q = \overline{A}.I_1 + A.I_0$$

We can increase the number of data inputs to be selected further simply by following the same procedure and larger multiplexer circuits can be implemented using smaller 2-to-1 multiplexers as their basic building blocks. So for a 4-input multiplexer we would therefore require two data select lines as 4-inputs represents 2^2 data control lines give a circuit with four inputs, I_0, I_1, I_2, I_3 and two data select lines A and B as shown.

4-to-1 Channel Multiplexer

The Boolean expression for this 4-to-1 Multiplexer above with inputs A to D and data select lines a, b is given as:

$$Q = \overline{a}\overline{b}A + a\overline{b}B + \overline{a}bC + abD$$

In this example at any one instant in time only ONE of the four analogue switches is closed, connecting only one of the input lines A to D to the single output at Q. As to which switch is closed depends upon the addressing input code on lines "a" and "b".

So for this example to select input B to the output at Q, the binary input address would need to be "a" = logic "1" and "b" = logic "0". Thus we can show the selection of the data through the multiplexer as a function of the data select bits as shown.

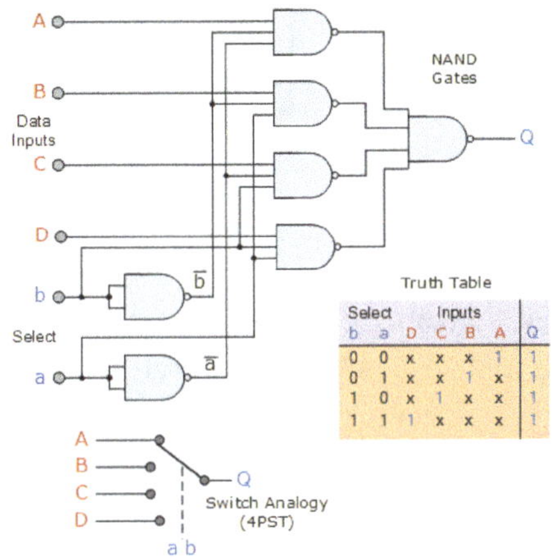

Select		Inputs				Q
b	a	D	C	B	A	Q
0	0	x	x	x	1	1
0	1	x	x	1	x	1
1	0	x	1	x	x	1
1	1	1	x	x	x	1

Multiplexer Input Line Selection

Adding more control address lines, (n) will allow the multiplexer to control more inputs as it can switch 2n inputs but each control line configuration will connect only ONE input to the output.

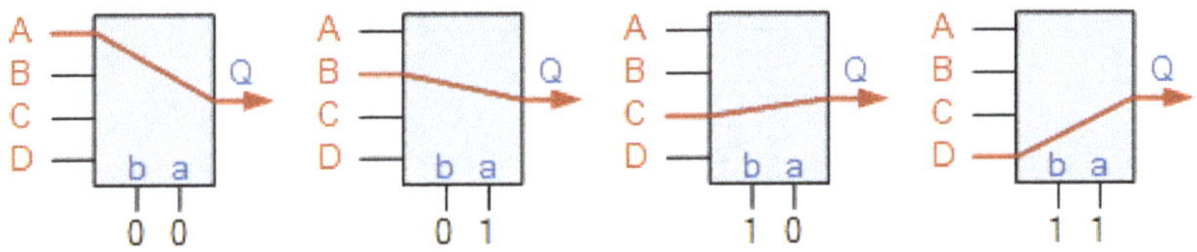

Then the implementation of the Boolean expression above using individual logic gates would require the use of seven individual gates consisting of AND, OR and NOT gates as shown.

4 Channel Multiplexer using Logic Gates

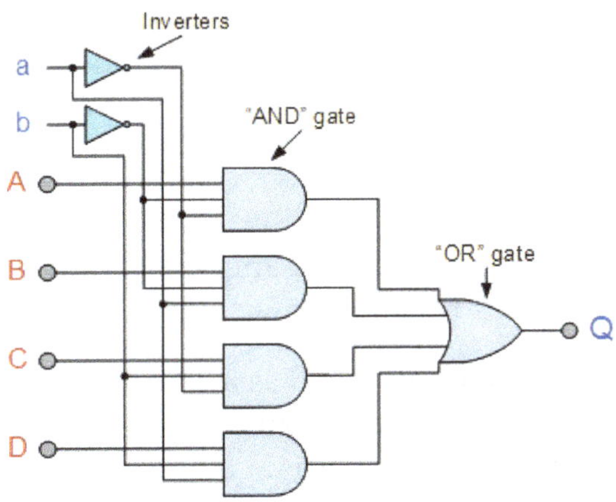

The symbol used in logic diagrams to identify a multiplexer is as follows:

Multiplexer Symbol

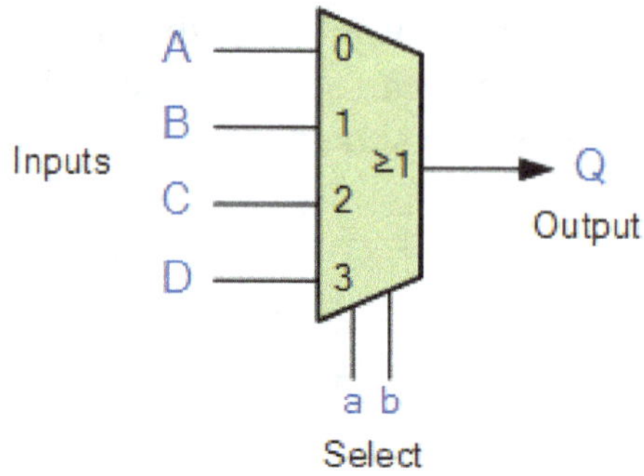

Multiplexers are not limited to just switching a number of different input lines or channels to one common single output. There are also types that can switch their inputs to multiple outputs and have arrangements or 4-to-2, 8-to-3 or even 16-to-4 etc configurations and an example of a simple Dual channel 4 input multiplexer (4-to-2) is given below:

4-to-2 Channel Multiplexer

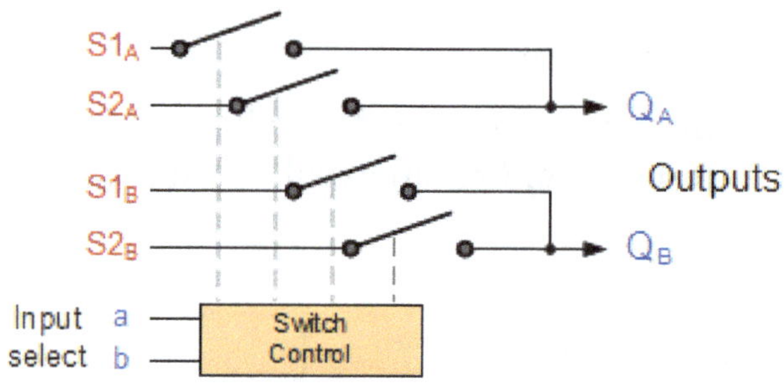

Here in this example the 4 input channels are switched to 2 individual output lines but larger arrangements are also possible. This simple 4-to-2 configuration could be used for example, to switch audio signals for stereo pre-amplifiers or mixers.

Adjustable Amplifier Gain

As well as sending parallel data in a serial format down a single transmission line or connection, another possible use of multi-channel multiplexers is in digital audio applications as mixers or where the gain of an analogue amplifier can be controlled digitally, for example.

Digitally Adjustable Amplifier Gain

Here, the voltage gain of the inverting operational amplifier is dependent upon the ratio between the input resistor, R_{IN} and its feedback resistor, Rf.

A single 4-channel (Quad) SPST switch configured as a 4-to-1 channel multiplexer is connected in series with the resistors to select any feedback resistor to vary the value of Rf. The combination of these resistors will determine the overall voltage gain of the amplifier, (Av). Then the voltage gain of the amplifier can be adjusted digitally by simply selecting the appropriate resistor combination.

Digital multiplexers are sometimes also referred to as "Data Selectors" as they select the data to be sent to the output line and are commonly used in communications or high speed network switching circuits such as LAN´s and Ethernet applications.

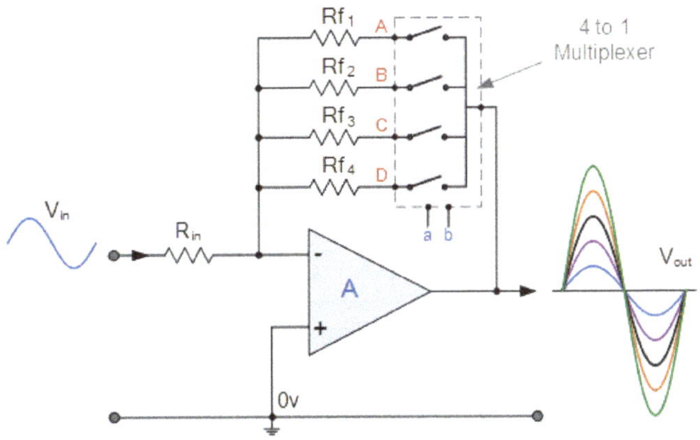

Some multiplexer IC´s have a single inverting buffer (NOT Gate) connected to the output to give a positive logic output (logic "1", HIGH) on one terminal and a complimentary negative logic output (logic "0", LOW) on another different terminal.

Demultiplexer (Demux)

The process of getting information from one input and transmitting the same over one of many outputs is called demultiplexing. A demultiplexer is a combinational logic circuit that receives the information on a single input and transmits the same information over one of 2n possible output lines.

The bit combinations of the select lines control the selection of specific output line to be connected to the input at given instant. The below figure illustrates the basic idea of demultiplexer, in which the switching of the input to any one of the four outputs is possible at a given instant.

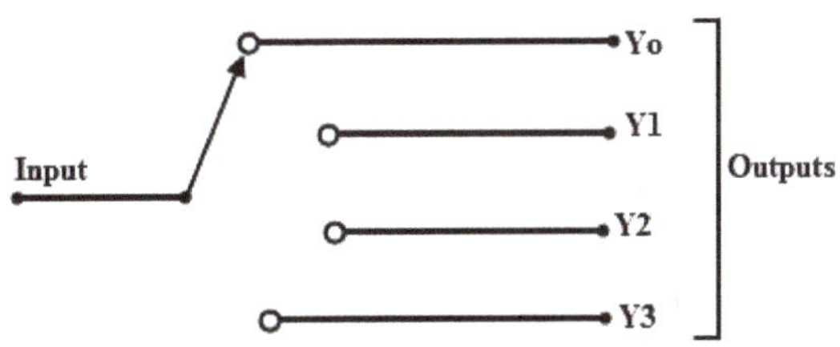

Demultiplexers are also called as data distributors, since they transmit the same data which is received at the input to different destinations.

Thus, a demultiplexer is a 1-to-N device where as the multiplexer is an N-to-1 device. The figure below shows the block diagram of a demultiplexer or simply a DEMUX.

It consists of 1 input line, n output lines and m select lines. In this, m selection lines are required to produce 2m possible output lines (consider 2m = n). For example, a 1-to-4 demultiplexer requires 2 (22) select lines to control the 4 output lines.

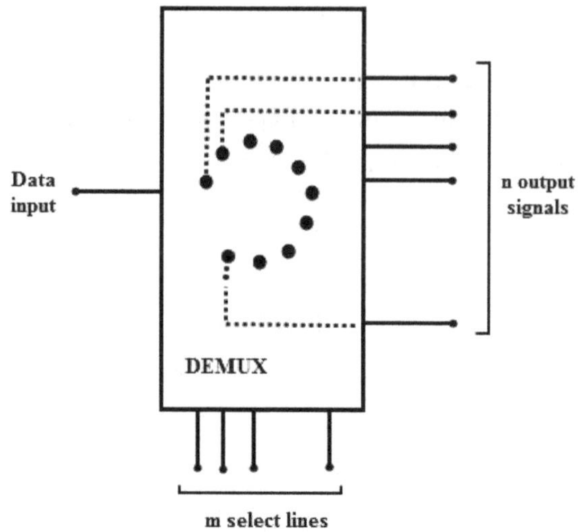

There are several types of demultiplexers based on the output configurations such as 1:4, 1:8 and 1:16.

These are available in different IC packages and some of the most commonly used demultiplexer ICs includes 74139 (dual 1:4 DEMUX), 73136 (1:8 DEMUX), 74154 (1:16 DEMUX), 74159 (1:16 DEMUX open collector type), etc.

1-to-2 Demultiplexer

A 1-to-2 demultiplexer consists of one input line, two output lines and one select line. The signal on the select line helps to switch the input to one of the two outputs. The figure below shows the block diagram of a 1-to-2 demultiplexer with additional enable input.

In the figure, there are only two possible ways to connect the input to output lines, thus only one select signal is enough to do the demultiplexing operation. When the select input is low, then the input will be passed to Y0 and if the select input is high then the input will be passed to Y1.

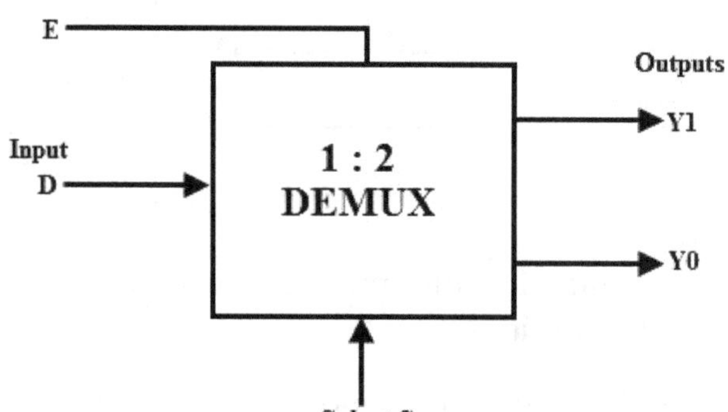

The truth table of a 1-to-2 demultiplexer is shown below in which the input is routed to Y0 and Y1 depends on the value of select input S. In the table output Y1 is active when the combination of select line and input line are active high, i.e., S F = 11.

Therefore, the output $Y_1 = SF$ and similarly the output Yo is equal to $\bar{S}$ F

Select	Input	Outputs	
S	D	Y_1	Y_0
0	0	0	0
0	1	0	1
1	0	0	0
1	1	1	0

From the above truth table, the logic diagram of this demultiplexer can be designed by using two AND gates and one NOT gate as shown in below figure. When the select lines S=0, AND gate A1 is enabled while A2 is disabled.

Then, the data from the input flows to the output line Y1. Similarly, when S=1, AND gate A2 is enabled and AND gate A1 is disabled, thus data is passed to the Yo output.

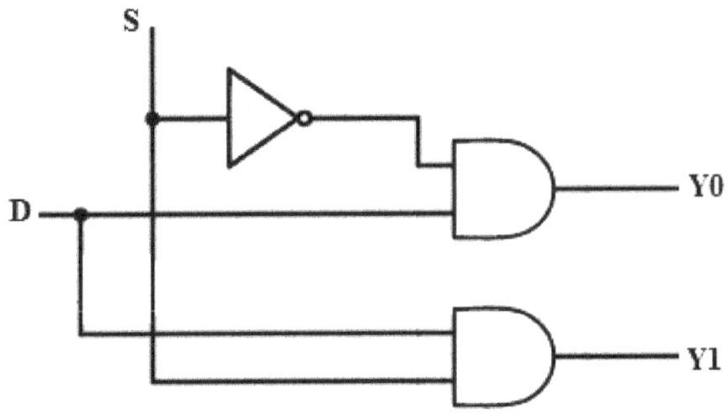

1-to-4 Demultiplexer

A 1-to-4 demultiplexer has a single input (D), two selection lines (S1 and So) and four outputs (Yo to Y3). The input data goes to any one of the four outputs at a given time for a particular combination of select lines.

This demultiplexer is also called as a 2-to-4 demultiplexer which means that two select lines and 4 output lines. The block diagram of 1:4 DEMUX is shown below.

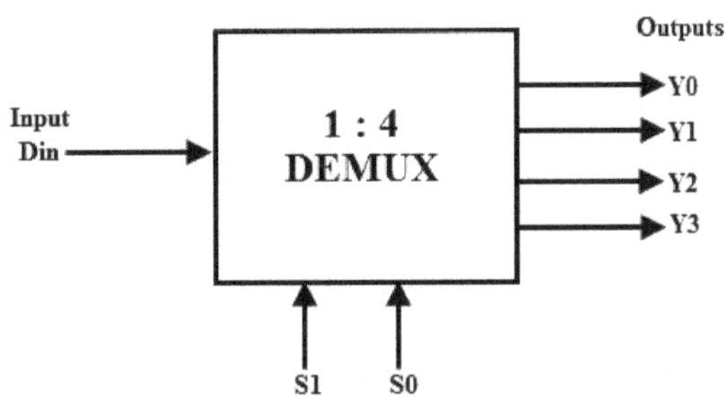

The truth table of this type of demultiplexer is given below. From the truth table it is clear that, when S1=0 and S0= 0, the data input is connected to output Y0 and when S1= 0 and s0=1, then the data input is connected to output Y1.

Similarly, other outputs are connected to the input for other two combinations of select lines.

Data Input	Select Inputs		Outputs			
D	S_1	S_0	Y_3	Y_2	Y_1	Y_0
D	0	0	0	0	0	D
D	0	1	0	0	D	0
D	1	0	0	D	0	0
D	1	1	D	0	0	0

From the table, the output logic can be expressed as min terms and are given below:

$$Y0=\overline{S1}\,\overline{S0}\,D$$

$$Y1=\overline{S1}\,S0\,D$$

$$Y2=S1\,\overline{S0}\,D$$

$$Y3=S1\,S0\,D$$

Where D is the input data, Y0 to Y3 are output lines and S0 & S1 are select lines.

From the above Boolean expressions, a 1-to-4 demultiplexer can be implemented by using four 3-input AND gates and two NOT gates as shown in figure below. The two selection lines enable the particular gate at a time.

So depends on the combination of select inputs, input data is passed through the selected gate to the associated output.

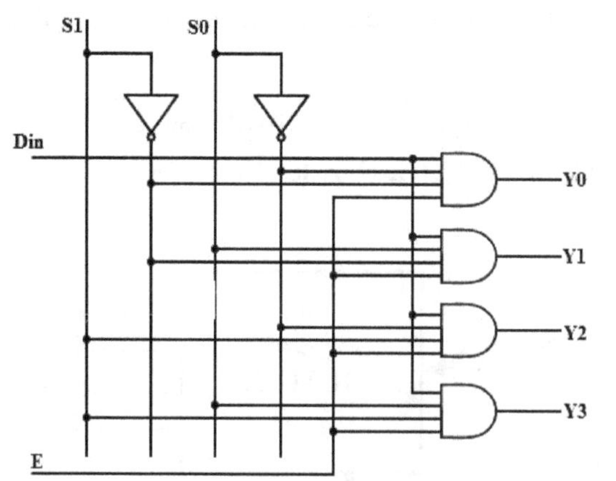

This type of demultiplexer is available in IC form and a typical IC 74139 is most commonly used dual 1-to-4 demultiplexer. It has two independent demultiplexers and each DEMUX accepts two binary inputs as select lines and four mutually exclusive active-low outputs.

Both demultiplexers share a common set of selection lines so they are selected in parallel. Also, each demultiplexer consists of enable pin or data input, for one demultiplexer it is active high data input and for other it is active low data input.

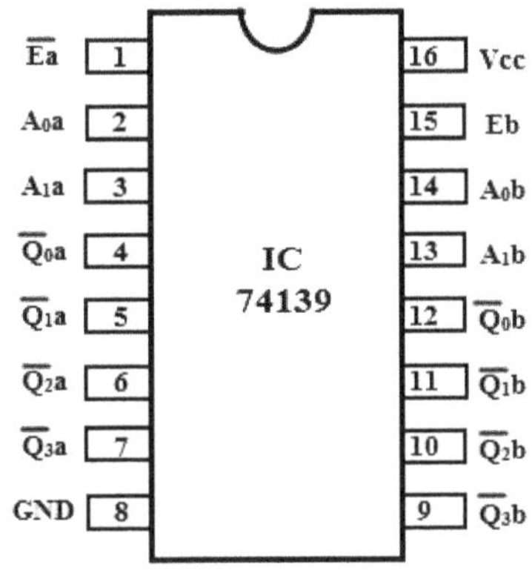

1-to-8 Demultiplexer

The below figure shows the block diagram of a 1-to-8 demultiplexer that consists of single input D, three select inputs S2, S1 and S0 and eight outputs from Y0 to Y7.

It is also called as 3-to-8 demultiplexer due to three select input lines. It distributes one input line to one of 8 output lines depending on the combination of select inputs.

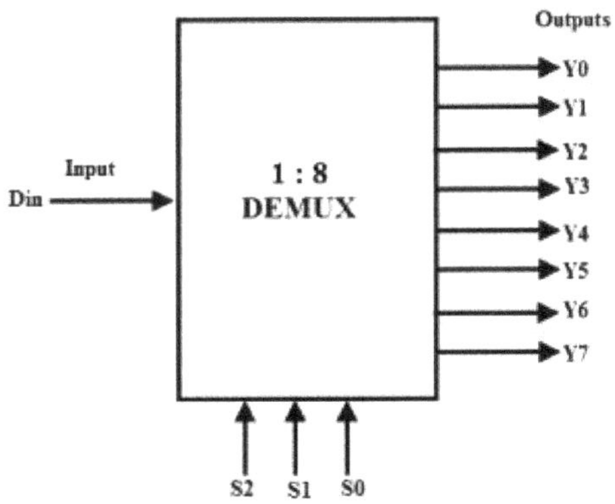

The truth table for this type of demultiplexer is shown below. The input D is connected with one of the eight outputs from Y0 to Y7 based on the select lines S2, S1 and S0.

For example, if S2S1S0=000, then the input D is connected to the output Y0 and so on.

Data Input	Select Inputs			Outputs							
D	S_2	S_1	S_0	Y_7	Y_6	Y_5	Y_4	Y_3	Y_2	Y_1	Y_0
D	0	0	0	0	0	0	0	0	0	0	D
D	0	0	1	0	0	0	0	0	0	D	0
D	0	1	0	0	0	0	0	0	D	0	0
D	0	1	1	0	0	0	0	D	0	0	0
D	1	0	0	0	0	0	D	0	0	0	0

D	1	0	1	0	0	D	0	0	0	0	0
D	1	1	0	0	D	0	0	0	0	0	0
D	1	1	1	D	0	0	0	0	0	0	0

From this truth table, the Boolean expressions for all the outputs can be written as follows:

$$Y0=D\overline{S2}\,\overline{S1}\,\overline{S0}$$

$$Y1=D\overline{S2}\,\overline{S1}S0$$

$$Y2=D\overline{S2}S1\overline{S0}$$

$$Y3=D\overline{S2}S1S0$$

$$Y4=DS2\overline{S1}\,\overline{S0}$$

$$Y5=DS2\overline{S1}S0$$

$$Y6=DS2S1\overline{S0}$$

$$Y7=DS2S1S0$$

From these obtained equations, the logic diagram of this demultiplexer can be implemented by using eight AND gates and three NOT gates as shown in below figure. The different combinations of the select lines, select one AND gate at given time, such that data input will appear at a particular output.

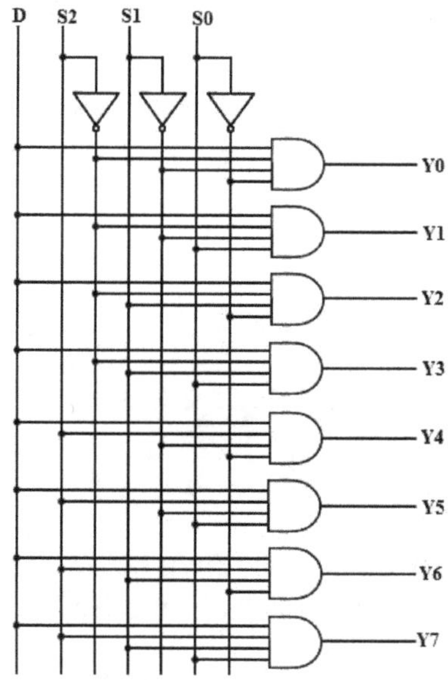

A typical IC74237 is a 1-to-8 demultiplexer that consists of latches at three select inputs. The pin out of this IC is given below.

The pins A0 to A2 are data inputs, Y0 to Y7 are demultiplexer outputs, E1&E2 are active-low data enable and active-high data enable pins respectively, LE is the latch enable input,Vcc and GND terminals are positive supply voltage and ground terminals.

With a 3-bit storage latch, this IC combines the 3-to-8 decoder function.

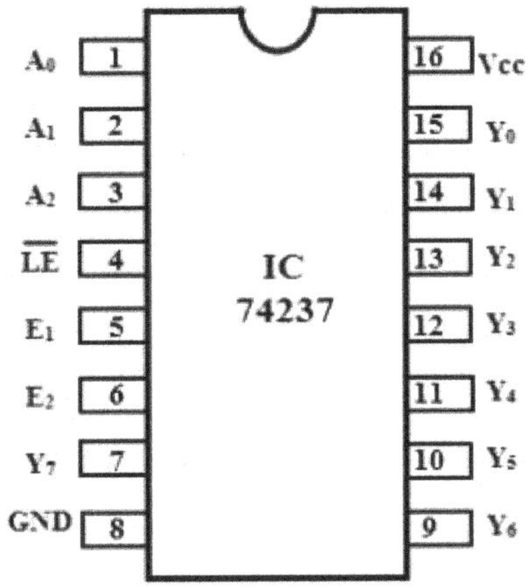

1-to-8 DEMUX using Two 1-to- 4 Demultiplexers

When the application requires a large demultiplexer with more number of output pins, then we cannot implement by a single integrated circuit. In case if more than 16 output pins are needed, then two or more demultiplexer ICs are cascaded to fulfill the requirement.

For example, if the application needs 32 output lines from a DEMUX, then we cascade two 1:16 demultiplexers or three 1:8 demultiplexers. Therefore, by cascading the two or more demultiplexers, a large demultiplexer can be implemented.

Consider the case that a 1-to-8 demultiplexer can be implemented by using two 1-to-4 demultiplexers with a proper cascading.

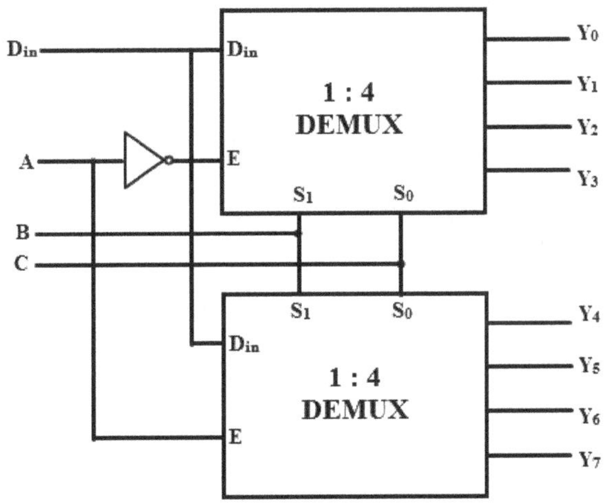

In the above figure, the highest significant bit A of the selection inputs are connected to the enable inputs such that it is complemented before connecting to one DEMUX and to the other it is directly connected.

By this configuration, when A is set to zero, one of the output lines from Y0 to Y3 is selected based on the combination of select lines B and C. Similarly, when A is set to one, based on the select lines one of the output lines from Y4 to Y7 will be selected.

Implementation of Full Subtractor using 1-to-8 DEMUX

As similar to the multiplexers, demultiplexers are also used for Boolean function implementation as well as combinational circuit design. We can design the demultiplexer to produce any truth table output by correspondingly controlling the select lines.

Consider the case for implementing a demultiplexer circuit in order to produce the full subtractor output. The truth table below shows the output of a full subtractor.

A	B	B_{IN}	D	B_{OUT}
0	0	0	0	0
0	0	1	1	1
0	1	0	1	1
0	1	1	0	1
1	0	0	1	0
1	0	1	0	0
1	1	0	0	0
1	1	1	1	1

From the above table, the full subtractor output D can be written as:

$$D = f(A, B, C) \\ = \sum m\ (1, 2, 4, 7)$$

And the borrow output can be expressed as:

$$B_{out} = F(A,B,C) = \sum m(1,2,3,7)$$

From these Boolean functions, a demultiplexer for producing full subtractor output can be built by properly configuring the 1-to-8 DEMUX such that with input D=1 it gives the minterms at the output.

And by logically ORing these minterms, the outputs of difference and borrow can be obtained as shown in figure.

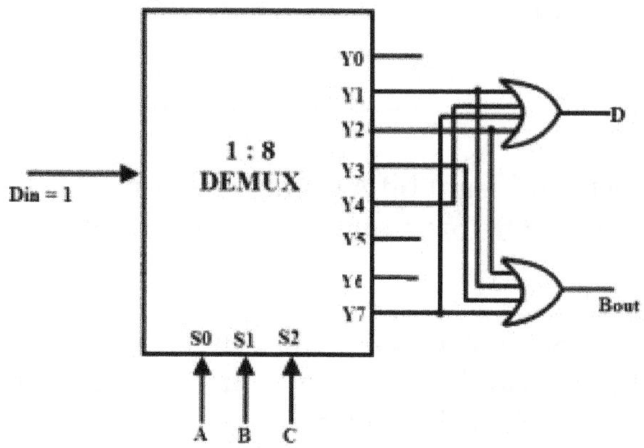

Applications of Demultiplexer

Since the demultiplexers are used to select or enable the one signal out of many, these are extensively used in microprocessor or computer control systems such as:

- Selecting different IO devices for data transfer.
- Choosing different banks of memory.
- Depends on the address, enabling different rows of memory chips.
- Enabling different functional units.

Other than these, demultiplexers can be found in a wide variety of application such as

- Synchronous data transmission systems.
- Boolean function implementation.
- Data acquisition systems.
- Combinational circuit design.
- Automatic test equipment systems.
- Security monitoring systems (for selecting a particular surveillance camera at a time), etc.

Sequential Circuits

Sequential logic circuits are those, whose output depends not only on the present value of the input but also on previous values of the input signal (history of values) which is in contrast to combinational circuits where output depends only on the present values of the input, at any instant of time. Sequential circuit can be considered as combinational circuit with feedback circuit. Sequential circuit uses a memory element like flip – flops as feedback circuit in order to store past values. The block diagram of a sequential logic is shown below.

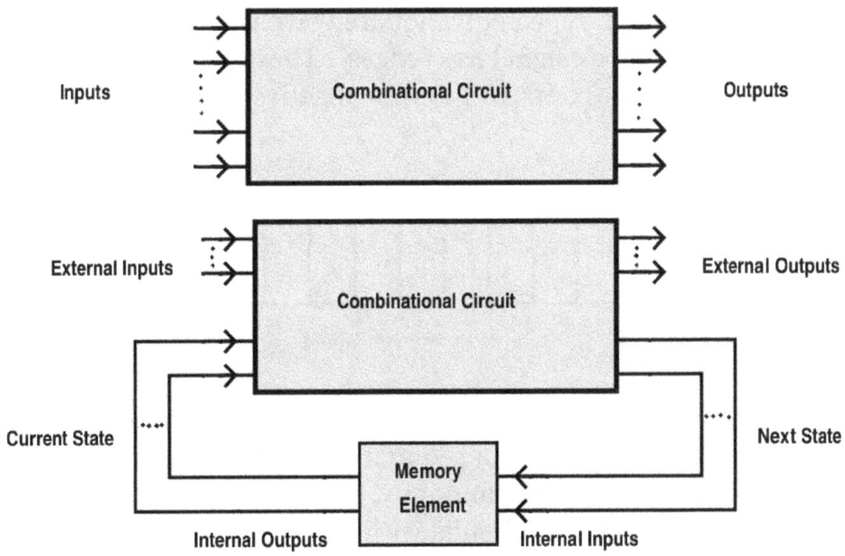

Sequential logic circuits are used to construct finite state machines, which are basic building block in all digital circuitry, and also in memory circuits. Basically, all circuits in practical digital devices are a mixture of combinational and sequential logic circuits.

Generally, we come across many counters in our daily life to count the number of objects. For example to count the number of audience entering or leaving an auditorium or to count number of vehicles in parking.In this when any person enters in to auditorium the counter increments its value depending on its present value. Similarly, it decrements its value depending on its previous and present value. So Counter retains the present state of the counter to do next operation.

This is analogous to sequential circuits which changes their state according to the previous and present signals.

Combinational Circuits vs Sequential Circuits

Combinational Circuits	Sequential Circuits
Output depends only on the present value of the inputs.	Output depends on both the present and previous state values of the inputs
These circuits will not have any memory as their outputs change with the change in the input value.	Sequential circuits have some sort of memory as their output changes according to the previous and present values.
There are no feedbacks involved.	In a sequential circuit the outputs are connected to it as a feedback path.
Used in basic Boolean operations.	Used in the designing of memory devices.
Implemented in: Half adder circuit, full adder circuit, multiplexers, demultiplexers, decoders and encoders.	Implemented in: RAM, Registers, counters and other state retaining machines.

Clock Signal in Sequential Circuits

The clock signal plays a crucial role in sequential circuits. A clock is a signal, which oscillates between logic level 0 and logic level 1, repeatedly. Square wave with constant frequency is the most common form of clock signal. A clock signal has "edges". These are the instants at which the clock changes from 0 to 1 (a positive edge) or from 1 to 0 (a negative edge).

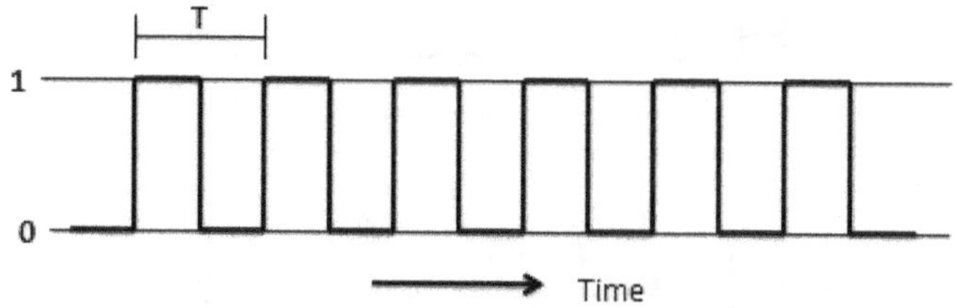

Clock signals control the outputs of the sequential circuit.That is it determines when and how the memory elements change their outputs. If a sequential circuit is not having any clock signal as input, the output of the circuit will change randomly. So that they cannot retain their state till the next input signal arrives. But sequential circuits with clock input will retain its state till the next clock edge occurs.

Classification of Sequential Circuits

Based on the clock signal input, the sequential circuits are classified into two types.

- Synchronous sequential circuit
- Asynchronous sequential circuit

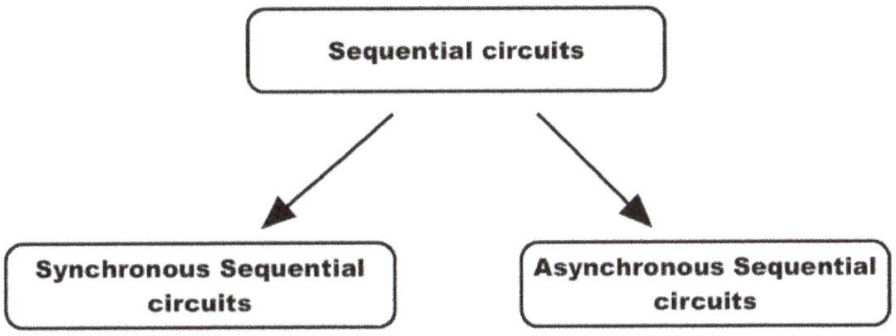

Synchronous Sequential Circuits

In Synchronous sequential circuit, the output depends on present and previous states of the inputs at the clocked instances. The circuits use a memory element to store the previous state. The memory elements in these circuits will have clocks. All these clock signals are driven by the same clock signal.

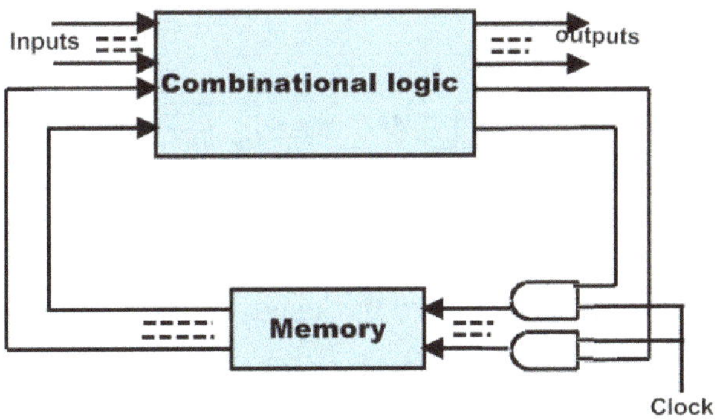

- Using clock signal, state changes will occur across all storage elements.
- These circuits are bit slower compared to asynchronous because they wait for the next clock pulse to arrive to perform the next operation.
- These circuits can be clocked or pulsed.
- The Synchronous sequential circuits that use clock pulses in their inputs are called clocked-sequential circuits. They are very stable.
- The sequential circuits that change their state using the pulse and these are called pulsed or un-clocked sequential circuits.
- Used in the design of MOORE-MEALY state management machines.
- They are used in synchronous counters, flip flops etc.

Limitations of Synchronous Sequential Circuits

- All the flip – flops in synchronous sequential circuits must be connected to clock signal. Clock signals are very high frequency signals and clock distribution consumes and dissipated a large amount of heat.
- Critical path or the slowest path determines the maximum possible clock frequency. Hence they are slower than asynchronous circuits.

Asynchronous Sequential Circuits

The Sequential circuits which do not operate by clock signals are called "Asynchronous sequential circuits".

- These circuits will change their state immediately when there is a change in the input signal.
- The Circuit behaviour is determined by signals at any instant in time and the order in which input signals change.

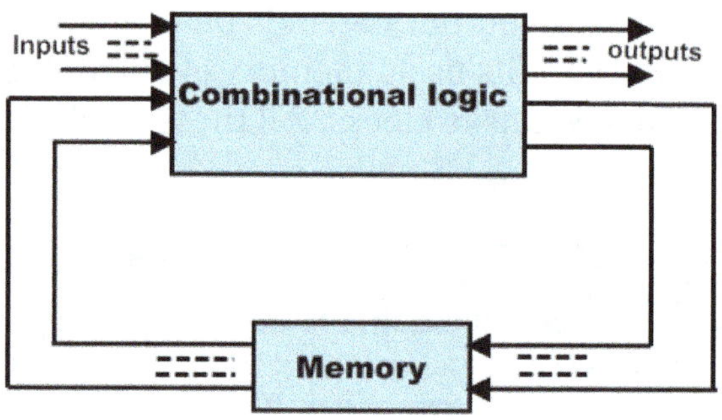

- They do not operate in pulse mode.
- They have better performance but hard to design due to timing problems.
- Mostly we use the asynchronous circuits when we require the low power operations.
- They are faster than synchronous sequential circuits as they do not need to wait for any clock signal.

Need of Asynchronous Sequential Circuits

These are used when speed of operation is important. As they are independent of internal clock pulse, they are operate quickly. so they are used in Quick response circuits.

- Used in the communication between two units having their own independent clocks.
- Used when we require the better external input handling.

Limitations of Asynchronous Sequential Circuits

- Asynchronous sequential circuits are more difficult to design.
- Though they have a faster performance, their output is uncertain.

Feedback in Sequential Circuits

Combinational circuits do not require any feed back as the outputs are purely dependent on the present value of the input. But in case of sequential circuits, the outputs are dependent on past values of the input along with present values. In order to involve memory element like a flip – flop, feedback must be introduced in the circuit. For example, consider a simple feedback circuit as shown below.

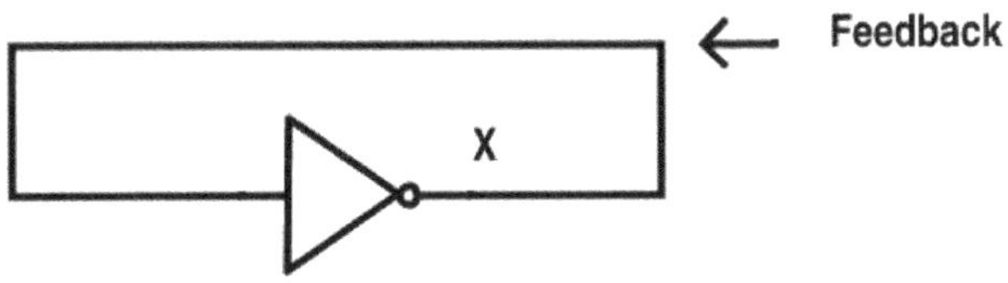

If 0 is the input to the inverter at an instance, this 0 will propagate and the output is 1. This 1 is fed back as input. This 1 will propagate and the output is 0. The process repeats and the result is a continuous oscillation of output between 0 and 1. There is no stable state in this scenario.

Now consider the following example of two inverters connected as shown.

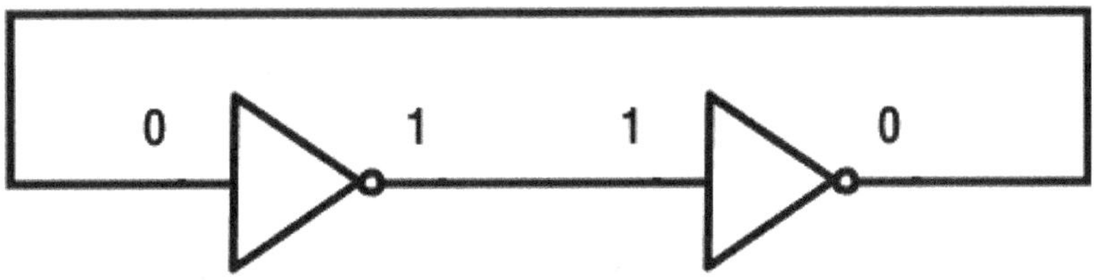

Here two inverters are connected back to back with the output of the second inverter fed back to input of first inverter. If 0 is the input to first inverter at an instance, it propagates through the first inverter and the output is 1. This 1 is input to second inverter and propagates through it. The output of the second inverter is 0 which is fed back to the first inverter. But the input to first inverter is already 0 and hence no change occurs. The circuit is said to be in a stable circuit. Another stable state can be obtained when the input to the first inverter is 1.

Flip-Flop

A flip flop is an electronic circuit with two stable states that can be used to store binary data. The stored data can be changed by applying varying inputs. Flip-flops and latches are fundamental building blocks of digital electronics systems used in computers, communications, and many other types of systems. Flip-flops and latches are used as data storage elements. It is the basic storage element in sequential logic. But first, let's clarify the difference between a latch and a flip-flop.

Flip Flop v/s Latch

The basic difference between a latch and a flip-flop is a gating or clocking mechanism.

For example, let us talk about SR latch and SR flip-flops. In this circuit when you Set S as active the output Q would be high and Q' will be low. This is irrespective of anything else. (This is an active-low circuit so active here means low, but for an active high circuit active would mean high).

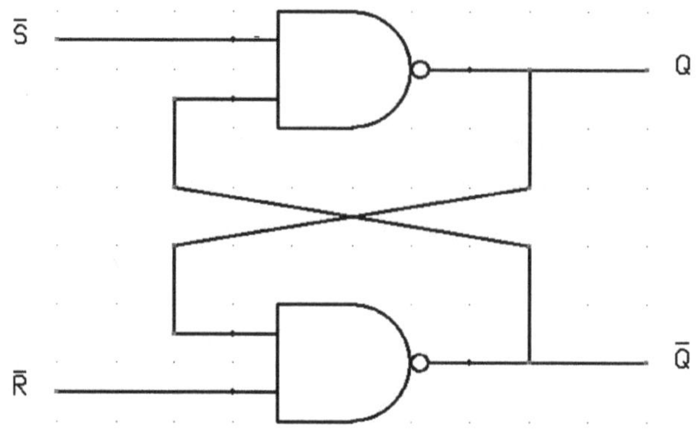

A flip flop, on the other hand, is synchronous and is also known as gated or clocked SR latch.

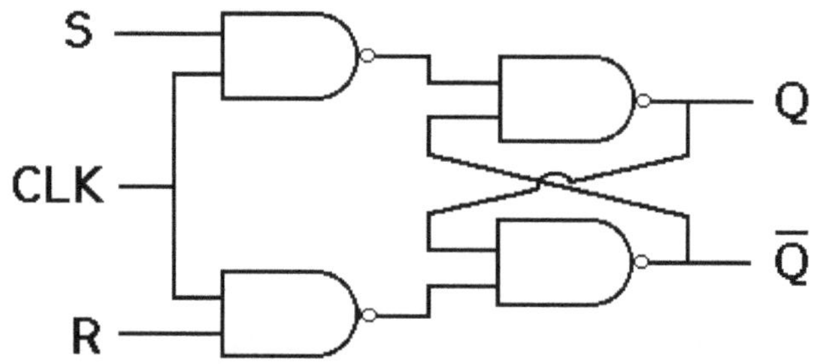

In this circuit diagram, the output is changed (i.e. the stored data is changed) only when you give an active clock signal. Otherwise, even if the S or R is active the data will not change. Let's look at the types of flip-flops to understand better.

SR Flip Flop

There are majorly 4 types of flip flops, with the most common one being SR flip flop. This simple flip flop circuit has a set input (S) and a reset input (R). In this circuit when you Set "S" as active the output "Q" would be high and "Q"" will be low. Once the outputs are established, the wiring of the circuit is maintained until "S" or "R" go high, or power is turned off. As shown above, it is the simplest and the easiest to understand. The two outputs, as shown above, are the inverse of each other. The truth table of SR Flip Flop is highlighted below.

S	R	Q	Q'
0	0	0	1
0	1	0	1
1	0	1	0
1	1	∞	∞

JK Flip-Flop

Due to the undefined state in the SR flip flop, another flip flop is required in electronics. The JK flip flop is an improvement on the SR flip flop where S=R=1 is not a problem.

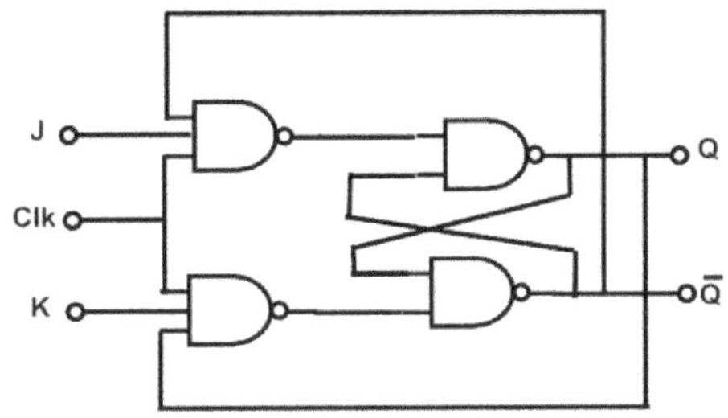

JK Flip-Flop

The input condition of J=K=1, gives an output inverting the output state. However, the outputs are the same when one tests the circuit practically.

In simple words, If J and K data input are different (i.e. high and low) then the output Q takes the value of J at the next clock edge. If J and K are both low then no change occurs. If J and K are both high at the clock edge then the output will toggle from one state to the other. JK Flip Flop can function as Set or Reset Flip flop.

J	K	Q	Q'
0	0	0	0
0	1	0	0
1	0	0	1
1	1	0	1
0	0	1	1
0	1	1	0
1	0	1	1
1	1	1	0

D Flip Flop

D flip flop is a better alternative that is very popular with digital electronics. They are commonly used for counters and shift-registers and input synchronisation.

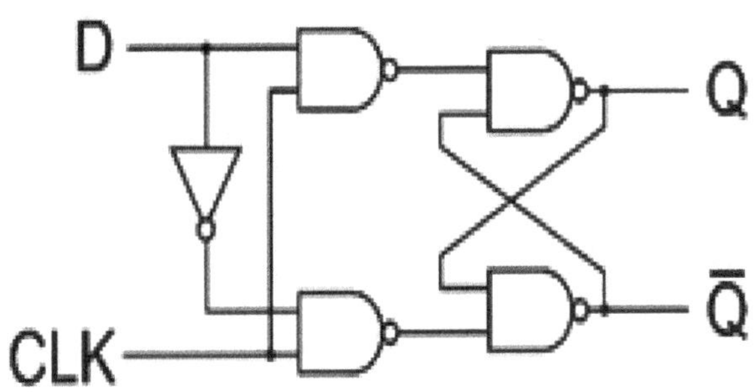

D Flip-Flop

In a D flip flop, the output can be only changed at the clock edge, and if the input changes at other times, the output will be unaffected.

Clock	D	Q	Q'
↓ » 0	0	0	1
↑ » 1	0	0	1
↓ » 0	1	0	1
↑ » 1	1	1	0

The change of state of the output is dependent on the rising edge of the clock. The output (Q) is same as the input and can only change at the rising edge of the clock.

T Flip Flop

A T flip flop is like JK flip-flop. These are basically a single input version of JK flip flop. This modified form of JK flip-flop is obtained by connecting both inputs J and K together. This flip-flop has only one input along with the clock input.

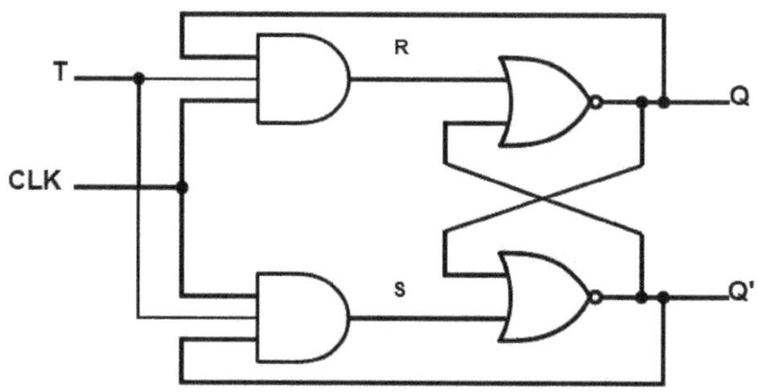

T Flip-Flop

These flip-flops are called T flip-flops because of their ability to complement its state (i.e.) Toggle, hence the name Toggle flip-flop.

T	Q	Q (t+1)
0	0	0
1	0	1
0	1	1
1	1	0

Applications of Flip-Flops

These are the various types of flip-flops being used in digital electronic circuits and the applications of Flip-flops are as specified below:

- Counters
- Frequency Dividers
- Shift Registers
- Storage Registers

Combinational and Sequential Circuits Comparison Chart

The main differences between difference between sequential and combinational circuit can be tabulated as below:

Basis of difference	Combinational Logic Circuits	Sequential Logic Circuits
Output	The output is in the form of functions of the current inputs or time independent logic.	The output is in the form of functions related to clock, current inputs as well as the previous states of a given system.
Storage of data and memory space allocation	Combinational circuits are not capable of storing data (state). They do not contain any memory element.	Sequential circuits have the memory space needed for storing the current states sent as control inputs (enable) for the next set of operations. They contain memory elements that store data in digital circuits.
Requirement of feedback	They do not need any feedback. Combinational circuits generally output the input as per the logic designed.	These circuits need feedback from the output to save inputs in the memory space for the next set of operations.
Main usage	Mainly used or the purpose of Boolean and Arithmetic operations. The circuits are represented with the help of Boolean algebra; they are further simplified with the usage of basic and universal logic gates.	Useful for data storage and therefore used in RAM.
Elementary building blocks	The elementary building blocks of combinational circuits are logic gates.	The elementary building units of sequential logic circuits are Flip flops (binary storage devices).
Dependency on clocks	Combinational circuits are independent of clock, therefore triggering is not important for their operations.	Sequential logic circuits are clocked and so they are triggered for operations with the help of electronic pulses. The flip-flops, if triggered, are termed as synchronous sequential circuits. The circuits that are not triggered are termed as asynchronous sequential circuits.
Example of circuits	Adder [1+0=1; the dependency is only on current inputs i.e., 1 and 0].	Counter [Previous O/P +1=Current O/P; the dependency is on current inputs and the previous state of inputs.]
Application	These circuits are used in association with encoders, adders, decoders, subtractors, multiplexers, etc.	Sequential circuits are used in case of flip-flops and latches. In these circuits, latches are considered to be the simplest elements used for retaining the earlier state. The latches are also known as flip-flops.
Definition	Combinational logic circuit is a digital logic circuit wherein the output is capable of being determined with the help of logic functions related to the current state inputs.	Sequential logic circuit is a digital logic circuit wherein the output is capable of being determined with the usage of the logic function related to current state inputs as well as the previous state inputs.
Storage of state within the circuit	Combinational circuits are not capable of storing any state within them.	Sequential circuit are capable of retaining their earlier state on the basis of current inputs as well as the earlier states.

Working Principle of Analog Circuits and Digital Circuits

The working of an analog circuits can be done with a normal waveform for converting, storing and amplifying the signals, while a digital circuit alters the wave forms into pulse signals. This article discusses about what is the difference between analog circuit and digital circuit?

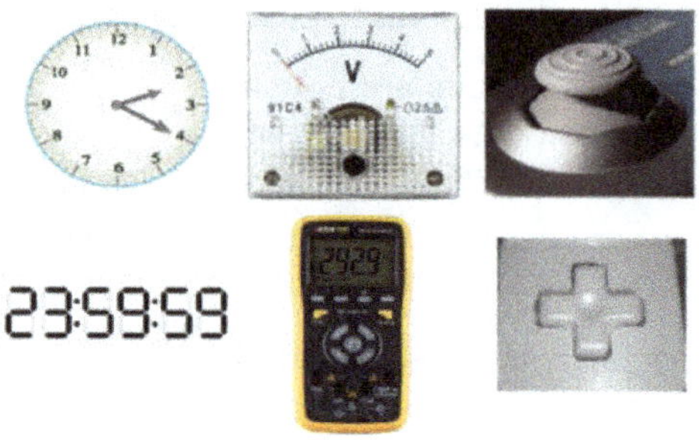

Analog Vs Digital

Difference between Analog Circuit and Digital Circuit

The main difference between analog circuit and digital circuit mainly includes what is an analog circuit, digital circuit and the differences between them:

Analog Circuit

The Analog electronic circuit includes an analog signal with any continuously changeable signal. While working on an analog signal, an analog circuit alters the signal in some manner. Analog circuit can be used to convert the original signal into some other format such as a digital signal. Analog circuits may also modify signals in inadvertent ways like adding noise or distortion. Analog circuits are classified into two types, namely active analog circuits and passive analog circuits. An analog circuit uses an electrical power source to get the goals of a designer while Passive circuits use no external electrical power.

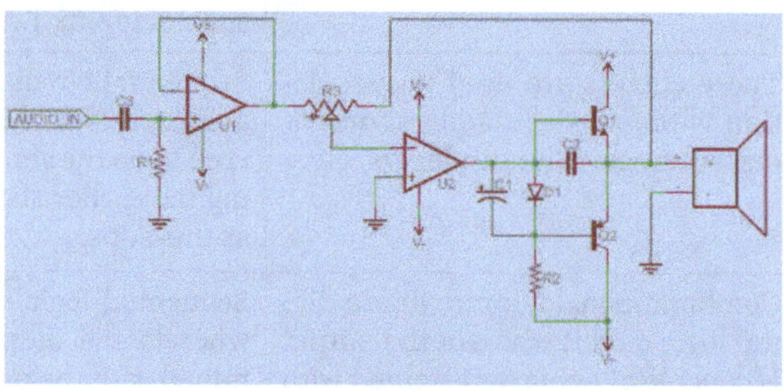

Analog Circuit

Digital Circuit

A digital circuit is a circuit where the signal should be one of two discrete levels. Each level is interpreted as one of two different states (for instance, 0 or 1). These circuits built with transistors to make logic gates in order to execute Boolean logic operation. This logic is the base of digital electronics & computer processing. Digital circuits are less vulnerable to degradation in excellence than analog circuits. It is also simpler to execute error detection and rectification with digital signals. To make the routine process of designing these circuits, designers use EDA (electronic design automation) tools, a kind of software that develops the logic in a digital circuit.

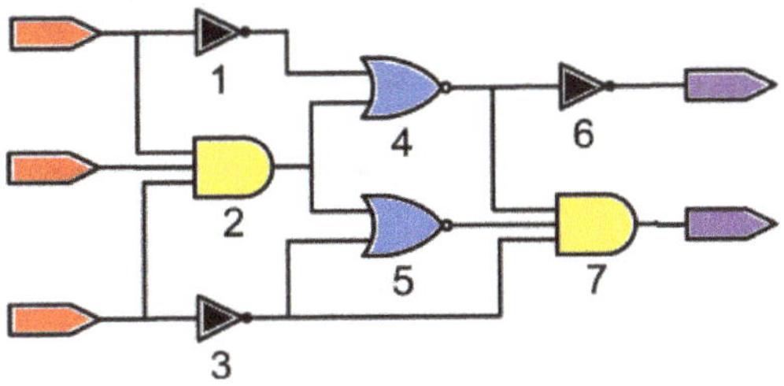

Digital Circuit

Working of Analog Circuit and Digital Circuits

The working of an analog circuits can be done with normal waveforms, changing them. For ex: a microphone is used in an analog circuit which converts the sound waves into analog electric waves. For example, in an analog circuit a microphone converts sound waves into similar or analog electric waves. These signals can be stored on the strip, improved in an analog amplifier and transformed back to related sound waves by a speaker.

A digital circuit is used to alter the waves into pulse waves. It measures a waveform thousand of times every second and stores the data in binary form. For instance, after 12msecs, a signal might be 2.4 volts high & after 14msecs, it might be at 2.6 volts. This circuit changes the volts and times into binary data, and the waves become a series of 1's and 0's. When the circuit has to generate sound from a speaker, it produces an o/p signal which is at 2.4 V after 12msec and at 2.6 V after 14msec, similar to the original wave.

O/P Quality

Analog circuit delivers an analog of the ordinary wave form and can generate very high o/p quality. Digital circuits generate an estimate of the natural waveform, so their quality of o/p is incomplete by the number of curve size they carry out.

Efficiency of a Circuit

An efficiency of a circuit mainly depends on how rapidly it can produce results and how much power it utilizes. These circuits have to wait for a wave to complete a cycle before they can carry out their tasks. And they require sufficient power to generate the peak of the wave. The digital circuit's speed is incomplete only by the speed of the gears that make up the circuit and not by the indications they are processing. To work with tiny pulses needs moderately little power. In most of the applications, analog circuits are slower & require more power than digital circuits.

Precision and Reproducibility

The act of analog circuits depends on their design and their gears to guarantee the waveform leftovers true to the original. They are liable to design faults, modifies from aging parts and outside powers such as electrical noise. Digital circuits just have to maintain path of their pulses. Even if numerous pulses get vanished, it influences only a few of the thousands of sizes.

Differences between Analog Circuits and Digital Circuits

At this point, the differences between Analog Circuits and Digital Circuits might be clear but as an aid for easy understanding, we have tabulated the differences.

Analog Circuits	Digital Circuits
Analog circuits operate on continuously variable signals also known as Analog Signals.	Digital Circuits operate on discretely variable signals or Digital signals i.e. the signal exists only in two levels: 0 and 1 (binary digital signalling).
Depending the efficiency and precision, it is quite difficult to design Analog Circuits.	Digital Circuits are relatively easy to design with many automated tools available for various stages of design and analysis.
When interacting with the physical world, analog circuits can directly accept the signals from outside as the data is already analog.	If a digital circuit has to acquire data from physical world, the analog signals must be converted to digital signals first.
As there is no need for data conversion, there is ideally no loss of information.	During the process of converting analog signals to digital signals, there might a significant amount of data loss, which can result in loss of information.
If precision and accuracy are not a criterion, then analog circuits can be simple and inexpensive.	Even with simple design techniques and at low cost, the digital circuits can provide good accuracy and precision.
Due to the lack of skilled engineers and the complexity of the designs, analog circuits can turnout to be quite expensive.	Advanced Integrated Circuits technologies and many other factors help the digital circuits to be reliable, lower in cost and smaller in size.

Integrated Circuit

Integrated circuit or IC or microchip or chip is a microscopic electronic circuit array formedby the fabrication of various electrical and electronic components (resistors, capacitors, transistors, and so on) on a semiconductor material (silicon) wafer, which can perform operations similar to the large discrete electronic circuits made of discrete electronic components.

As all these arrays of components, microscopic circuits and semiconductor wafer material base are integrated together to form a single chip, hence, it is called as integrated circuit or integrated chip or microchip.

Electronic circuits are developed using individual or discrete electronic components with different sizes, such that the cost and size of these discrete circuits increase with the number of components used in the circuit. To conquer this negative aspect, the integrated circuit technology was developed – Jack Kilby of Texas Instruments developed the first IC or integrated circuit in the 1950s and thereafter, Robert Noyce of Fairchild Semiconductor solved some practical problems of this-integrated circuit.

Integrated Circuits

Different Types of Integrated Circuits

There are different types of ICs; classification of Integrated Circuits is done based on various criteria. A few types of ICs in a system are shown in the below figure with their names in a tree format.

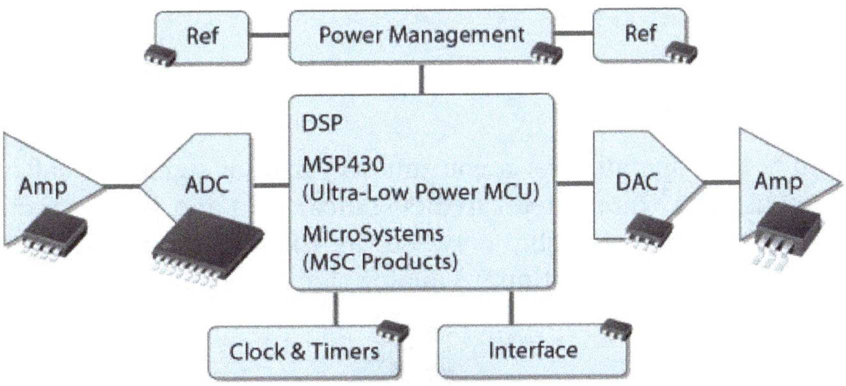

Different Types of ICs

Based on the intended application, the IC are classified as analog integrated circuits, digital integrated circuits and mixed integrated circuits.

Digital Integrated Circuits

The integrated circuits that operate only at a few defined levels instead of operating over all levels of signal amplitude are called as Digital ICs and these are designed by using multiple number of digital logic gates, multiplexers, flip flops and other electronic components of circuits.These logic gates work with binary input data or digital input data, such as 0 (low or false or logic 0) and 1 (high or true or logic 1).

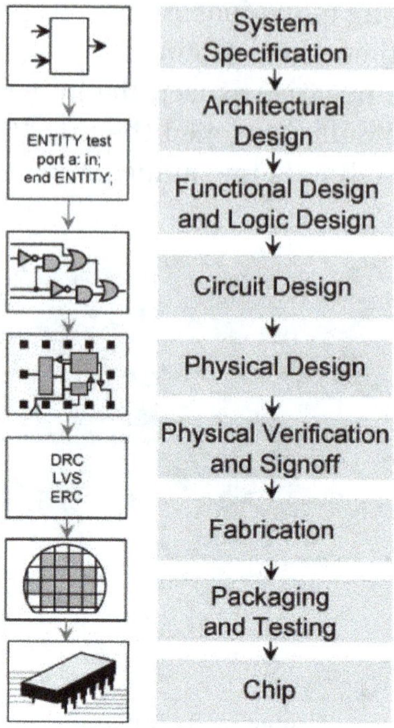

Digital Integrated Circuits

The above figure shows the steps involved in designing a typical digital integrated circuits.These digital ICs are frequently used in the computers, microprocessors, digital signal processors, computer networks and frequency counters. There are different types of digital ICs or types of digital integrated circuits, such as programmable ICs, memory chips, logic ICs, power management ICs and interface ICs.

Analog Integrated Circuits

The integrated circuits that operate over a continuous range of signal are called as Analog ICs. These are subdivided as linear Integrated Circuits (Linear ICs) and Radio Frequency Integrated Circuits (RF ICs). In fact, the relationship between the voltage and current maybe nonlinear in some cases over a long range of the continuous analog signal.

Analog Integrated Circuits

The frequently used analog IC is an operational amplifier or simply called as an op-amp, similar to the differential amplifier, but possesses a very high voltage gain. It consists of very less number of transistors compared to the digital ICs, and, for developing analog application specific integrated circuits (analog ASICs), computerized simulation tools are used.

Mixed Integrated Circuits

The integrated circuits that are obtained by the combination of analog and digital ICs on a single chip are called as Mixed ICs. These ICs functions as Digital to Analog converters, Analog to Digital converters (D/A and A/D converters) and clock/timing ICs. The circuit depicted in the above figure is an example of mixed integrated circuit which is a photograph of the 8 to 18 GHz self healing radar receiver.

Mixed Integrated Circuits

This mixed-signal Systems-on-a-chip is a result of advances in the integration technology, which enabled to integrate digital, multiple analog and RF functions on a single chip.

General types of integrated circuits (ICs) include the following:

Logic Circuits

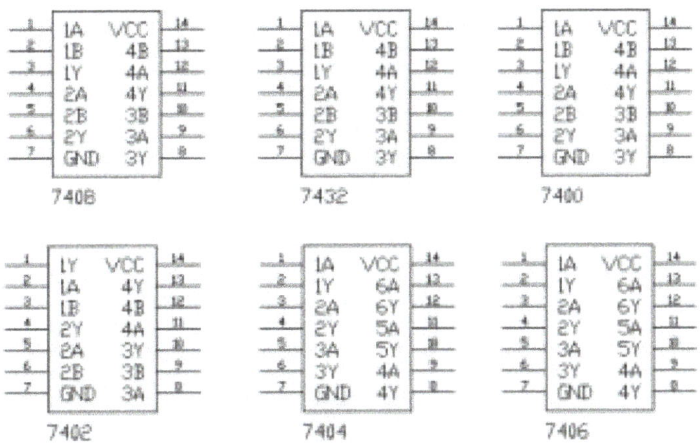

Logic Circuits

These ICs are designed using logic gates-that work with binary input and output (0 or 1). These are mostly used as decision makers. Based on the logic or truth table of the logic gates, all the logic gates connected in the IC give an output based on the circuit connected inside the IC- such that this output is used for performing a specific intended task. A few logic ICs are shown above.

Comparators

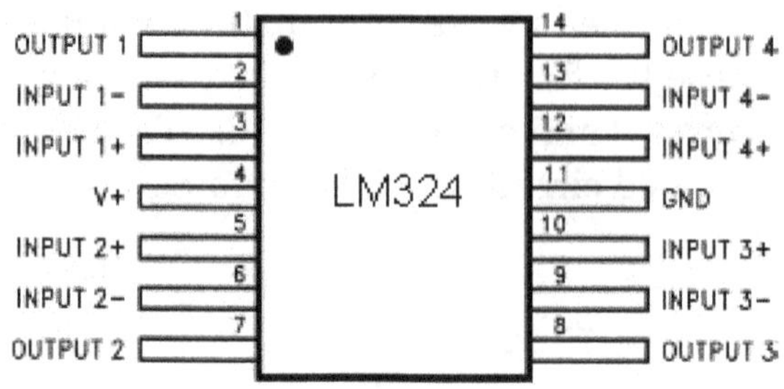

Comparators

The comparator ICs are used as comparators for comparing the inputs and then to produce an output based on the ICs' comparison.

Switching ICs

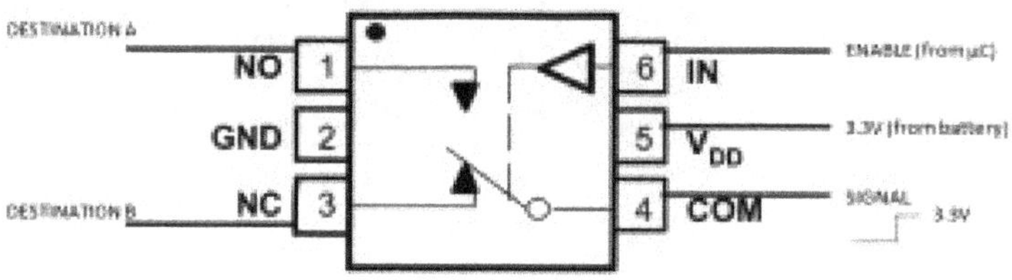

Switching ICs

Switches or Switching ICs are designed by using the transistors and are used for performing the switching operations. The above figure is an example showing an SPDT IC switch.

Audio Amplifiers

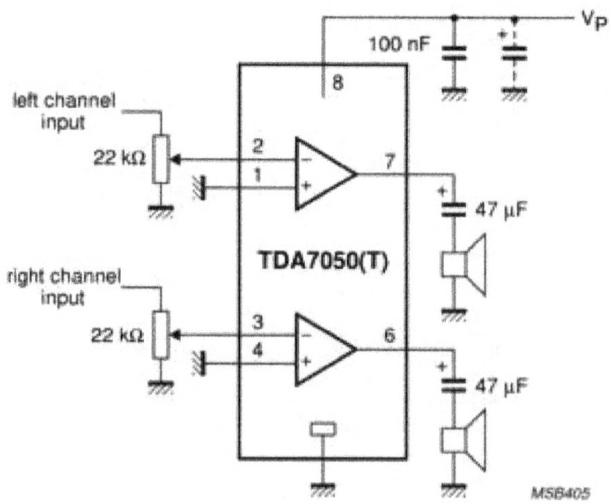

Audio amplifiers

The audio amplifiers are one of the many types of ICs, which are used for the amplification of the audio. These are generally used in the audio speakers, television circuits, and so on. The above circuit shows the low- voltage audio amplifier IC.

Operational Amplifiers

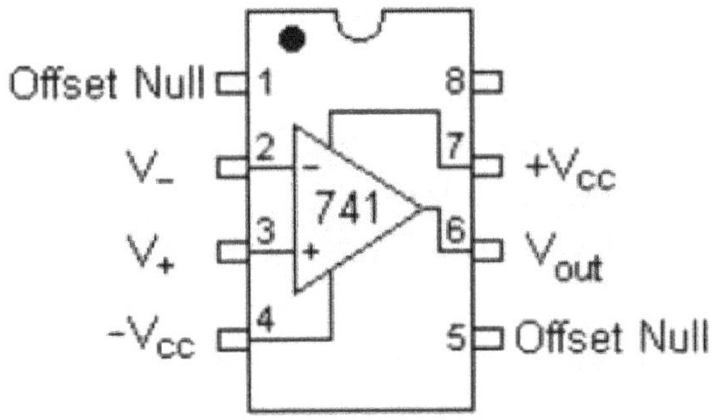

Operational amplifiers

The operational amplifiers are frequently used ICs, similar to the audio amplifiers which are used for the audio amplification. These op-amps are used for the amplification purpose, and these ICs work similar to the transistor amplifier circuits. The pin configuration of the 741 op-amp IC is shown in the above figure.

Timer ICs

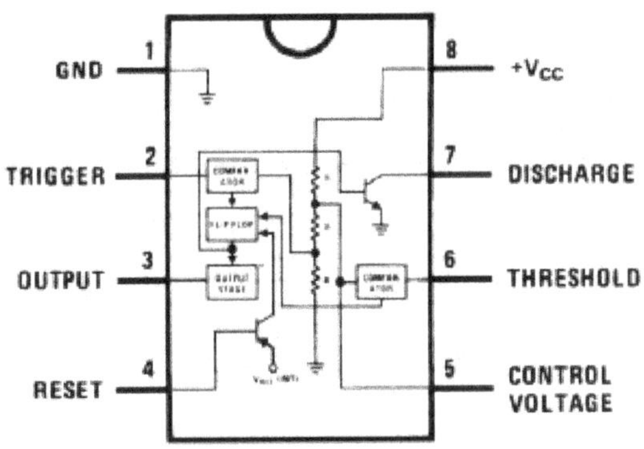

LM555 Timer

Timer ICs

Timers are special purpose integrated circuits used for the purpose of counting and to keep a track of time in intended applications. The block diagram of the internal circuit of the LM555 timer IC is shown in the circuit.

Based on the number of components used (typically based on the number of transistors used), they are as follows:

- Small-scale integration consists of only a few transistors (tens of transistors on a chip), these ICs played a critical role in early aerospace projects.
- Medium-scale integration consists of some hundreds of transistors on the IC chip developed in the 1960s and achieved better economy and advantages compared to the SSI ICs.

- Large-scale integration consists of thousands of transistors on the chip with almost the same economy as medium scale integration ICs. The first microprocessor, calculator chips and RAMs of 1Kbit developed in the 1970s had below four thousand transistors.
- Very large-scale integration consists of transistors from hundreds to several billions in number.
- Ultra large-scale integration consists of transistors in excess of more than one million, and later wafer-scale integration (WSI), system on a chip (SoC) and three dimensional integrated circuit (3D-IC) were developed.

References

- What-is-an-electronic-circuit, components, electronics, programming: dummies.com, Retrieved 8 June, 2019
- Analog-circuits-and-digital-circuits: electronicshub.org, Retrieved 18 August, 2019
- Analog-circuits, engineering: sciencedirect.com, Retrieved 27 March, 2019
- Basic-analog-circuits, white-papers, innovations: ni.com, Retrieved 9 January, 2019
- Analog-circuits-and-digital-circuits: electronicshub.org, Retrieved 14 April, 2019
- Digital-logic-circuits: edgefx.in, Retrieved 19 July, 2019
- Introduction-to-combinational-logic-circuits: electronicshub.org, Retrieved 30 June, 2019
- Binary-adder-half-and-full-adder: electrical4u.com, Retrieved 2 April, 2019
- Binary-subtractor, combination: electronics-tutorials.ws, Retrieved 29 August, 2019
- Demultiplexerdemux: electronicshub.org, Retrieved 13 May, 2019
- Sequential-circuits-basics: electronicshub.org, Retrieved 15 February, 2019
- Flip-flop, learn-electronics, resources: electronicsforu.com, Retrieved 6 June, 2019
- Difference-between-analog-circuit-and-digital-circuit: elprocus.com, Retrieved 23 January, 2019
- Different-types-of-integrated-circuits: elprocus.com, Retrieved 3 March, 2019

5

Analog and Digital Electronics

Electronic systems with a continuous variable signal are known as analog electronics. It includes operational amplifiers, analog multipliers, instrumentation amplifiers, etc. Digital electronics are the electronic systems that majorly operate on digital signals. It is made up of large assemblies of logic gates. The topics elaborated in this chapter will help in gaining a better perspective about the analog and digital electronics.

Analog Electronics

Analog electronics is a branch of electronics that deals with a continuously variable signal. It's widely used in radio and audio equipment along with other applications where signals are derived from analog sensors before being converted into digital signals for subsequent storage and processing. Although digital circuits are considered as a dominant part of today's technological world, some of the most fundamental components in a digital system are actually analog in nature.

Analog

Analog means continuous and real. The world we live in is analog in nature, implying that it's full of infinite possibilities. The number of smells we can sense, the number of tones we can hear, or the number of colors we can paint with; everything is infinite. The people working in the field of analog electronics are basically dealing with analog devices and circuits.

For example, if we build a circuit and it counts values like 1, 2, 3, 4, and 5; the values are neither infinite nor continuous. On the other hand, if the circuit counts like 1.00000, 1.00001, 1.00002, 4.99999, 5.00000, the amount of information would be infinite.

Analog Signals

In electrical engineering, signals are basically time-varying quantities (usually voltage or current). So when we talk about the signal it means we are talking about a voltage that's changing over time.

Signals are passed among devices in order to obtain or send information in the form of audio, video or encoded data. The transmission takes place through wires or via air through radio frequency waves. For instance, audio signals are transferred from the computer's audio card to the speakers, while data signals between a tablet and a Wi-Fi router pass through the air.

Analog Signals use attributes of the medium to convey the signal's information. For example, an aneroid barometer makes use of the angular position of a needle to convey the changes in atmospheric pressure. The signals take any value from a given range and each signal value denotes different information. Each level of the signal signifies a different level of the phenomenon and any change in the signal is meaningful.

It's quite easy to determine if a signal graph is analog or digital. The former one is smooth and continuous while the latter is edge and appears in the form of stepping squares. Given below is an analog signal graph representing the change in voltage with the change in time.

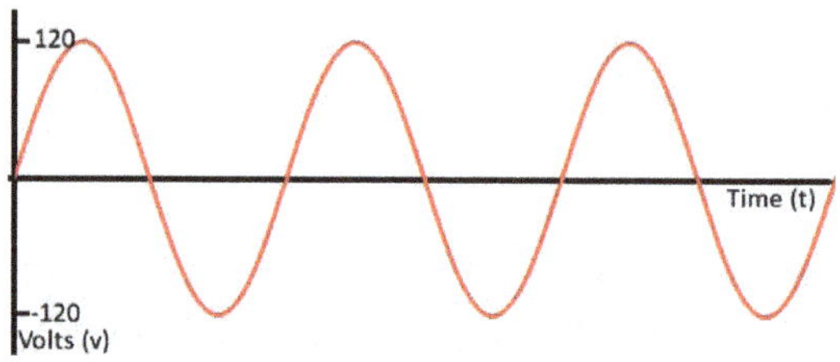

An analog signal graph representing time-voltage graph

Operational Amplifiers

Operational Amplifiers (OAs) are highly stable, high gain dc difference amplifiers. Since there is no capacitive coupling between their various amplifying stages, they can handle signals from zero frequency (dc signals) up to a few hundred kHz. Their name is derived by the fact that they are used for performing *mathematical operations* on their input signal(s).

Figure shows the symbol for an OA. There are two inputs, the *inverting input* (-) and the *non-inverting input* (+). These symbols have nothing to do with the polarity of the applied input signals.

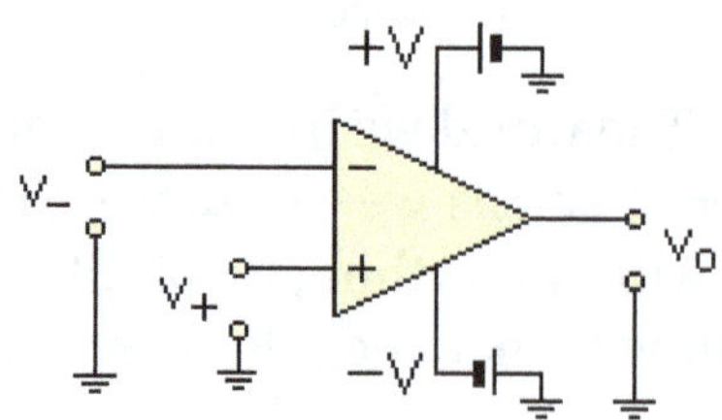

Figure: Symbol of the operational amplifier. Connections to power supplies are also shown

The output signal (voltage), v_o, is given by:

$$v_o = A(v_+ - v_-)$$

v_+ and v_- are the signals applied to the non-inverting and to the inverting input, respectively. *A* represents the open loop gain of the OA. A is infinite for the ideal amplifier, whereas for the various types of real OAs, it is usually within the range of 10^4 to 10^6.

OAs require two power supplies to operate, supplying a positive voltage (+V) and a negative voltage (-V) with respect to circuit common. This bipolar power supply allows OAs to generate output signals (results) of either polarity. The output signal (v_o) range is not unlimited. The voltages of the power supplies determine its actual range. Thus, a typical OA fed with -15 and +15 V, may yield a v_o within the (approximately) -13 to +13 V range, called operational range. Any result expected to be outside this range is clipped to the respective limit, and OA is in a saturation stage.

The connections to the power supplies and to the circuit common symbols, shown in figure, hereafter will be implied, and they will be not shown in the rest of the circuits for simplicity.

Because of their very high open loop gain, OAs are almost exclusively used with some additional circuitry (mostly with resistors and capacitors), required to ensure a negative feedback loop. Through this loop a tiny fraction of the output signal is fed back to the inverting input. The negative feedback stabilizes the output within the operational range and provides a much smaller but precisely controlled gain, the so-called closed loop gain.

Circuits of OAs have been used in the past as analog computers, and they are still in use for mathematical operations and modification of the input signals in real time. A large variety of OAs is commercially available in the form of low cost integrated circuits.

There is a plethora of circuits with OAs performing various mathematical operations. Each circuit is characterized by its own transfer function, i.e. the mathematical equation describing the output signal (v_o) as a function of the input signal (v_i) or signals (v_1, v_2, ..., v_n). Generally, transfer functions can be derived by applying Kirchhoff's rules and the following two simplifying assumptions:

1. The output signal (v_o) acquires a value that (through the feedback circuits) practically equates the voltages applied to both inputs, i.e. $v_+ \approx v_-$.
2. The input resistance of both OA inputs is extremely high(usually within the range 10^6-10^{12} MΩ, for the ideal OA this is infinite), thus no current flows into them.

Inverting Amplifier

The basic circuit of the inverting amplifier is shown in figure.

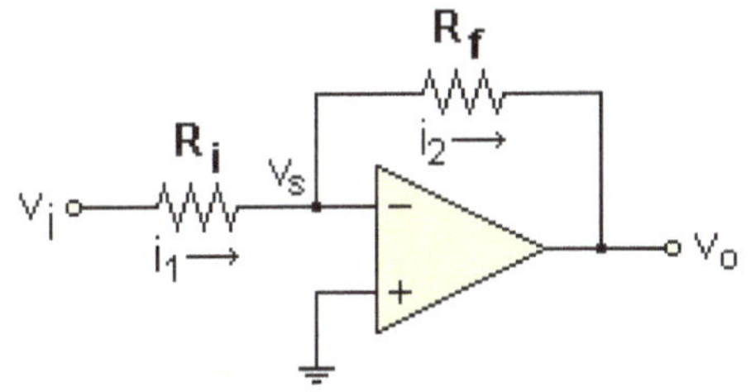

Figure: Inverting amplifier

The transfer function is derived as follows: Considering the arbitrary current directions we have:

$$i_1 = (v_i - v_s)/R_i \text{ and } i_2 = (v_s - v_o)/R_f$$

The non-inverting input is connected directly to the circuit common (i.e. $v_+ = 0$ V), therefore (considering simplifying assumption 1) $v_s = v_- = 0$ V, therefore:

$$i_1 = v_i/R_i \text{ and } i_2 = - v_o/R_f$$

Since there is no current flow to any input (simplifying assumption 2), it is:

$$i_1 = i_2$$

Therefore, the transfer function of the inverting amplifier is:

$$v_o = -(R_f/R_i)v_i$$

Thus, the closed loop gain of the inverting amplifier is equal to the ratio of R_f (feedback resistor) over R_i (input resistor). This transfer function describes accurately the output signal as long as the closed loop gain is much smaller than the open loop gain A of the OA used (e.g. it must not exceed 1000), and the expected values of v_o are within the operational range of the OA.

Summing Amplifier

The summing amplifier is a logical extension of the previously described circuit, with two or more inputs. Its circuit is shown in Figure.

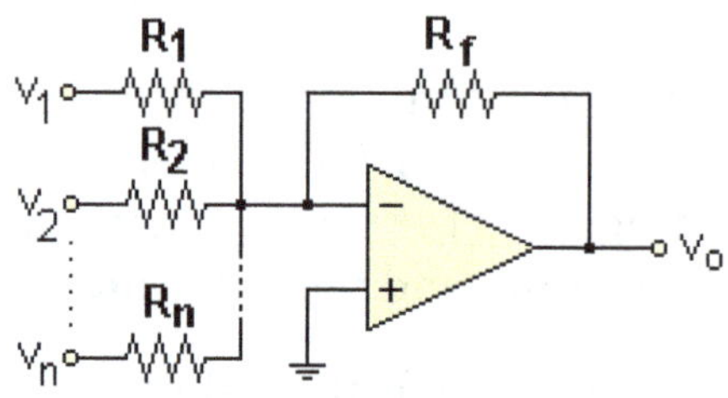

Figure: Summing amplifier

The transfer function of the summing amplifier (similarly derived) is:

$$v_o = -(v_1/R_1 + v_2/R_2 + \ldots + v_n/R_n)R_f$$

Thus if all input resistors are equal, the output is a scaled sum of all inputs, whereas, if they are different, the output is a weighted linear sum of all inputs.

The summing amplifier is used for combining several signals. The most common use of a summing amplifier with two inputs is the amplification of a signal combined with a subtraction of a constant amount from it (dc offset).

Difference Amplifier

Difference amplifier precisely amplifies the difference of two input signals. Its typical circuit is shown in Figure.

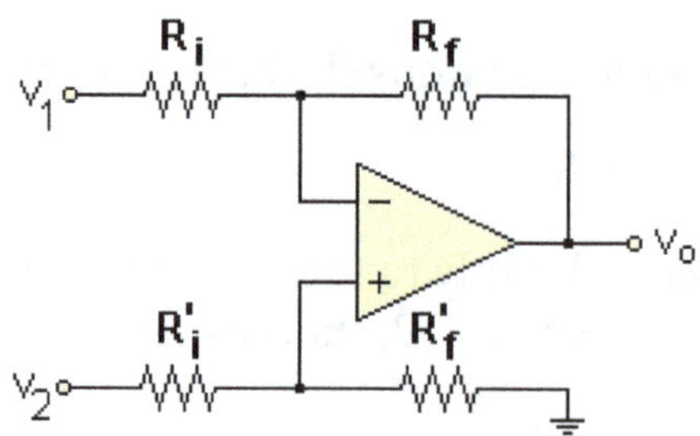

Figure: *Difference amplifier*

If $R_i = R_i'$ *and* $R_f = R_f'$, then the transfer function of the difference amplifier is:

$$v_o = (v_2 - v_1)\, R_f/R_i$$

The difference amplifier is useful for handling signals referring not to the circuit common, but to other signals, known as floating signal *sources. Its capability to reject a common signal makes it* particularly valuable for amplifying small voltage differences contaminated with the same amount of noise (common signal).

In order for the difference amplifier to be able to reject a large common signal and to generate at the same time an output precisely proportional to the two signals difference, the two ratios $p = R_f/R_i$ *and* $q = R_f'/R_i'$ *must be precisely equal*, otherwise the signal output will be:

$$v_o = [q(p+1)/(q+1)]v_2 - pv_1$$

Differentiator

The differentiator *generates an output signal proportional to the first derivative of the input with respect to time. Its typi*cal circuit is shown in Figure.

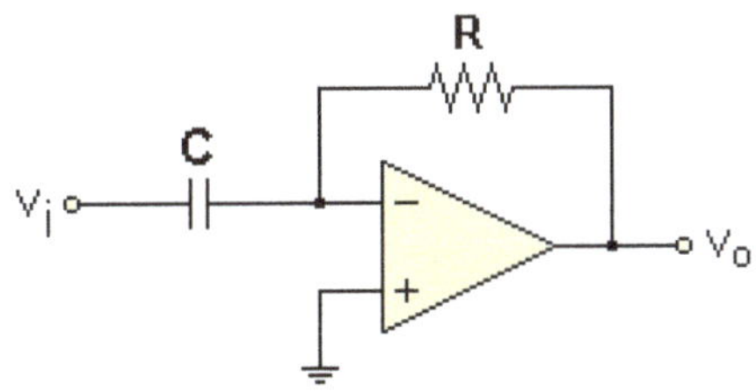

Figure: Differentiator

The transfer function of this circuit is:

$$v_o = -RC(dv_i/dt)$$

Obviously, a constant input (regardless of its magnitude) generates a zero output signal. A typical usage of the differentiator in the field of chemical instrumentation is obtaining the first derivative of a potentiometric titration curve for the easier location of the titration final points (points of maximum slope).

Integrator

The integrator generates an output signal proportional to the time integral of the input signal.Its typical circuit is shown in Figure.

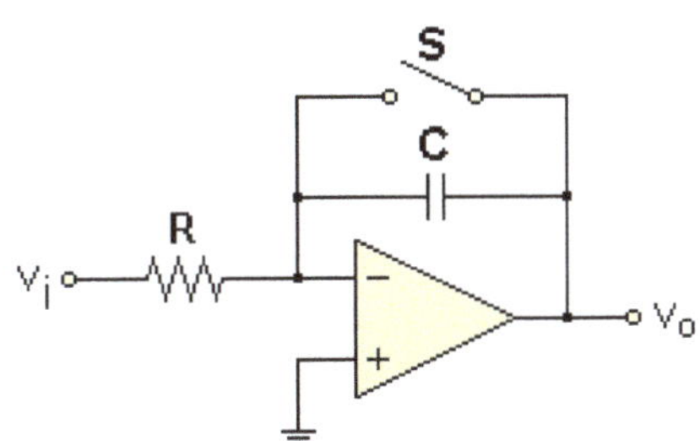

Figure: Integrator

$$v_o = -(1/RC)\int v_i(t)dt$$

The output remains zero as far as switch S remains closed. The integration starts (t = 0) when S opens. The output is proportional to the charge accumulated in capacitor C, which serves as the integrating device. A typical application of the (analog) integrator in chemical instrumentation is the integration of chromatographic peaks, since its output will be proportional to the peak area.

If the input signal is stable then the output from the integrator will be given by the equation:

$$v_o = -(v_i/RC)\, t$$

i.e. the output signal will be a voltage ramp. Voltage ramps are commonly used for generating the linear potential sweep required in polarography and many other voltammetric techniques.

Op-amp Comparator

A comparator, in electronics, is a circuit configuration that compares two voltages (or currents) and indicates which one is larger. Thus, the inputs to a comparator should be different in nature. Comparators can be easily configured using op-amps, since the op-amps have high gain and balanced difference inputs.

Theoretically, an op-amp in open-loop configuration (no feedback) can be used as a comparator. When the input voltage at the non-inverting terminal V_+ is greater than the voltage at the inverting input terminal V_-, the output of the op-amp saturates at its positive extreme. When the non-inverting input voltage drops below the inverting input voltage, the op-amp output switches to its negative saturation level. Comparator circuits are most widely used in analog-to-digital converters (ADCs) and in oscillators.

Op-amp Inverting Comparator

In an inverting comparator, the input voltage V_{in} is applied to the inverting input terminal of the op-amp and the non-inverting input terminal is connected to reference voltage, through resistors R_1 and R_2. As long as the input voltage V_{in} is lesser than the reference voltage V_{ref}, the output of the op-amp remains positively saturated. When V_{in} goes above the reference voltage, the output of the op-amp switches to its negative saturation level and remains negatively saturated as long as V_{in} is less than V_{ref}. The circuit of a comparator using op-amp is shown in the figure below.

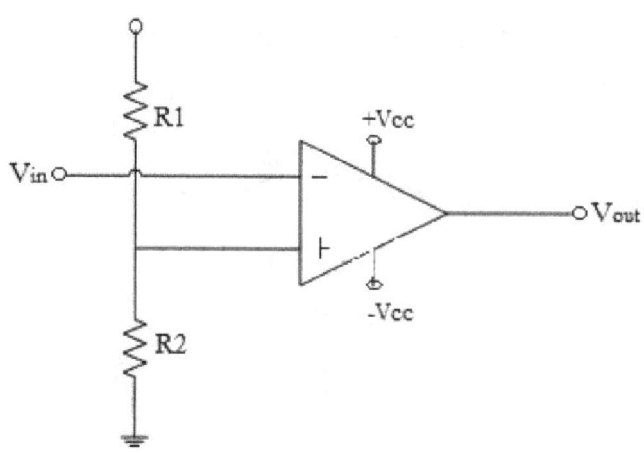

Figure: Inverting Comparator using Op-amp

By choosing the values of resistors R_1 and R_2, the reference voltage V_{ref} can be adjusted and comparator can be used to compare input voltage with the corresponding reference voltage.

$$V_{out} = +V_{sat};\ \text{if } V_{in} < V_{ref}$$
$$= -V_{sat};\ \text{if } V_{in} > V_{ref}$$

The input and output waveforms of an op-amp inverting comparator is shown in the figure below.

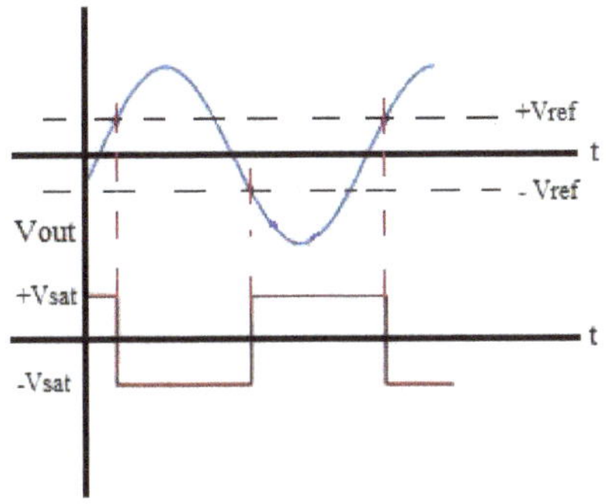

Figure: input output Waveforms of Inverting Comparator

Op-amp Non-Inverting Comparator

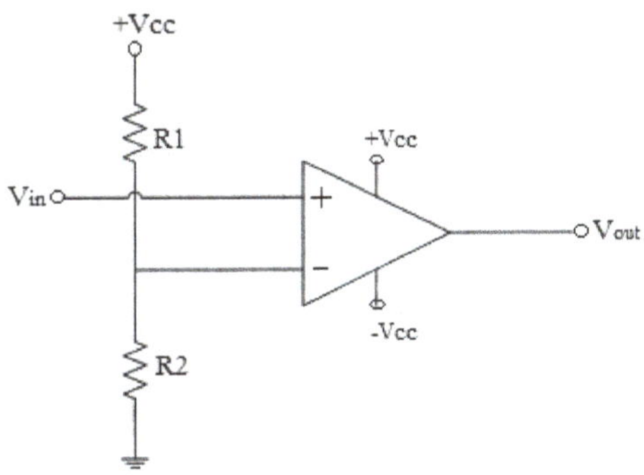

Figure: Non-Inverting using Op-amp

In the case of an op-amp non-inverting comparator, the input voltage V_{in} is applied to the non-inverting input terminal and the reference voltage, V_{ref}, is connected to the inverting input terminal. When the input voltage V_{in} is greater than the reference voltage V_{ref}, the op-amp output is positively saturated. In practice, the difference (V_{in}-V_{ref}) will be a positive value. Since there is no feedback to the op-amp input, the open-loop gain of the op-amp will be infinity. Hence the output will swing to its maximum possible value, +Vsat. When the input voltage falls below the reference voltage, the output switches to its negative saturation voltage.

$$V_{out} = +V_{sat};\ \text{if } V_{in} > V_{ref}$$
$$= -V_{sat};\ \text{if } V_{in} < V_{ref}$$

Op-amp Logarithmic Amplifier

An operational amplifier can be configured to function as a Logarithmic amplifier, or simply Log amplifier. Log amplifier is a non-linear circuit configuration, where the output is *K* times the logarithmic value of the input voltage applied. Log amplifiers find the applications in computations such as multiplication and division of signals, computation of powers and roots, signal compression and decompression, as well as in process control in industrial applications. A log amplifier can be constructed using a bipolar junction transistor in the feedback to the op-amp, since the collector current of a BJT is logarithmically related to its base-emitter voltage.

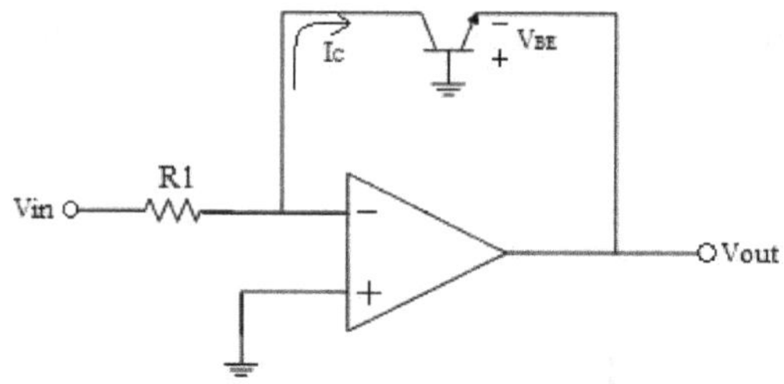

Figure: Op-amp Logarithmic Amplifier

The circuit of a fundamental log amplifier using op-amp is shown in the figure above. The necessary condition of the log amplifier to work is that the input voltage always must be positive. It can be seen that $V_{out} = -V_{be}$.

Since the collector terminal of the transistor is held at virtual ground and the base terminal is also grounded, the voltage-current relationship becomes that of a diode and is given by,

$$I_E = I_S.[e^{q(Vbe)/kT} - 1]$$

Where,

I_S = the saturation current,

k = Boltzmann's constant,

T = absolute temperature (in K).

Since, $I_E = I_C$ for grounded base transistor,

$$I_C = I_S.\ [e^{q(Vbe)/kT} - 1]$$

$$(I_C/I_S) = [e^{q(Vbe)/kT} - 1]$$

$$(I_C/I_S) + 1 = [e^{q(Vbe)/kT}]$$

$$(I_C+I_S)/I_S = e^{q(Vbe)/kT}$$

$$e^{q(Vbe)/kT} = (I_C/I_S) \text{ since } I_C >> I_S$$

Taking natural log on both sides of the above equation, we get:

$$V_{be} = (kT/q)\ \ln[I_C/I_S]$$

The collector current $I_C = V_{in}/R_1$ and $V_{out} = -V_{be}$

Therefore,

$$V_{out} = -(kT/q)\ \ln[V_{in}/R_1.I_S]$$

The output of the circuit is, thus, proportional to the log of the input voltage. However, the output is dependent on the saturation current which varies from transistor to transistor and also with temperature. Compensation circuits can be added to stabilize the output against these variations.

Anti-Logarithmic Amplifier or Exponential Amplifier

Anti-logarithmic or exponential amplifier (or simply antilog amplifier) is an op-amp circuit configuration, whose output is proportional to the exponential value or anti-log value of the input. Antilog amplifier does the exact opposite of a log amplifier. Antilog amplifiers along with log amplifiers are used to perform analogue computations on the input signals. The circuit of an antilog amplifier using op-amp is shown in the figure below.

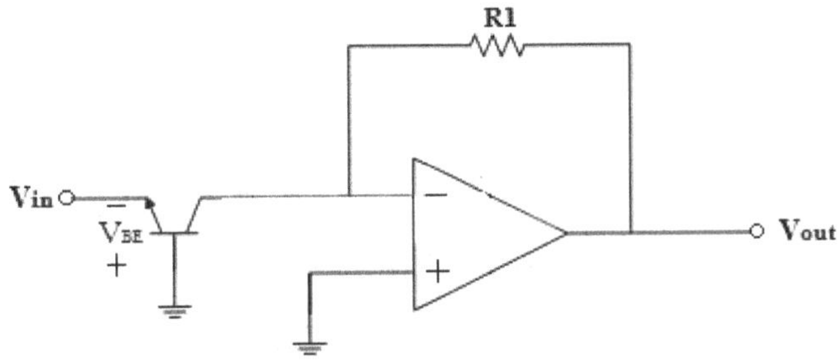

Figure: Anti-logarithmic Amplifier

It is noted that by exchanging the positions of the transistor and the resistor, the log amplifier can be made to work as antilog amplifier. The base-collector voltage of the transistor is maintained at ground potential, from the virtual ground concept. The current I_E for the transistor is given by,

$$I_E = I_S.[e^{q(Vbe)/kT} - 1]$$

For a grounded base transistor, $I_E = I_C$. Therefore,

$$I_C = I_S.[e^{q(Vbe)/kT} - 1]$$

Where, I_S = saturation current of the transistor,

$$V_{out} = I_C.R_1$$

$$V_{out} = I_S.[e^{q(Vbe)/kT} - 1].R_1$$

Also, for the above circuit $V_{in} = -V_{be}$. Therefore,

$$V_{out} = R_1.I_S.[e^{q(-Vin)}/kT - 1]$$

Antilog amplifiers also suffer from unstable outputs, due to the variations in I_S for different transistors and temperature dependence. Compensating circuits can be added to stabilize the output against such variations.

Current to Voltage Converter

An operational amplifier current-to-voltage converter, also known as a Trans-Impedance amplifier, is a circuit which converts the change in the input current to a corresponding output voltage. The current to voltage converter circuit is most commonly used to amplify the current output of photodiodes, photo-detectors, accelerometers and other sensor devices to a noticeable and usable voltage level.

A simple current-to-voltage converter circuit will have a feedback resistor, with a large value of resistance. The gain of the amplifier is dependent on this resistor. Depending on the application, a current-to-voltage converter can be constructed in different ways. All the configurations convert the low-level current output of a sensor device to a significant voltage level. The gain and the bandwidth of the circuit changes with different types of sensors.

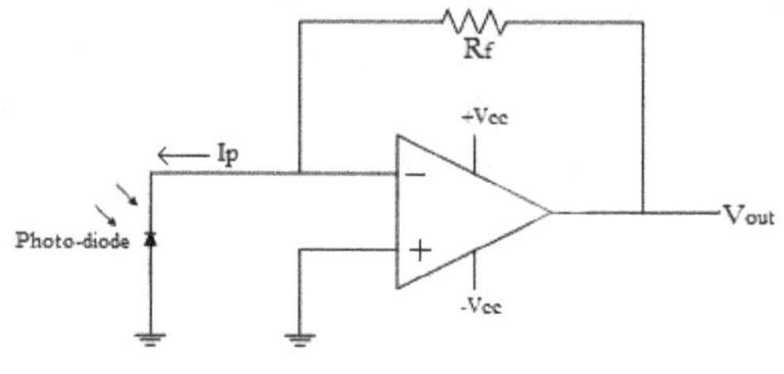

Figure: An op-amp Current-to-Voltage Converter

The circuit of a basic current-to-voltage converter is shown in the figure above. The sensor device, in this case is a photo-diode, and it is connected to the inverting input terminal and the non-inverting input terminal is grounded. This provides a low impedance load for the photo-diode, which keeps the voltage across the photo-diode low. The high gain of the op-amp keeps the photo-diode current, Ip, equal to the feedback current through resistor R_f. The input offset voltage due to the photo-diode is very low, since the photo-diode has no external bias. This provides a large output gain, without any output offset voltage.

The output voltage of the above circuit can be given as,

$$V_{out} = - I_p.R_f$$

The above equation satisfies only for the DC and low-frequency gain current-to-voltage converter. If the gain is large, any input offset voltage at the non-inverting input of the op-amp will result in an output offset voltage. To minimize these effects, the current-to-voltage converters are usually designed with FETs at the op-amp input, which have very low input offset voltages.

Op-amp Inverter

An op-amp inverting amplifier, or inverter, inverts the input signal as well as amplifies it. A positive-going signal at the input of an inverting amplifier would result in a negative-going signal at the output and vice versa.An AC sinusoidal signal at the input would produce 180° out of phase sinusoidal signal at the output.

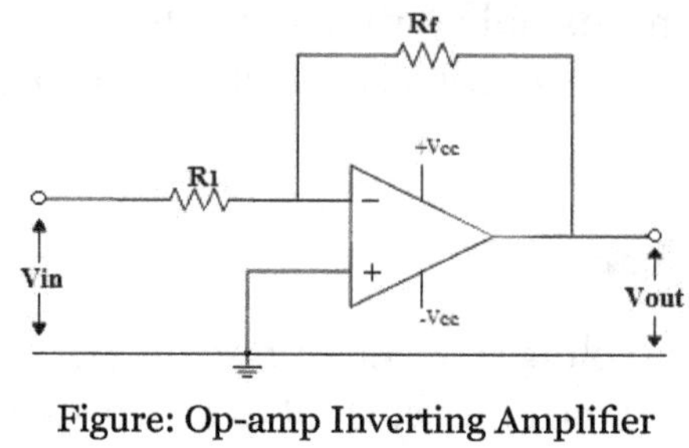

Figure: Op-amp Inverting Amplifier

Figure shows the circuit of a typical inverting amplifier using an op-amp. The circuit uses a negative feedback connection, through the resistor R_f. The input signal is applied to the inverting input terminal and the non-inverting input terminal is grounded.

Since the input current to the op-amp is ideally zero, the current I due to the input voltage flows through resistors R_1 and R_f. The input and output voltages can be calculated as,

$$V_{in} = I.R_1$$

$$V_{out} = -I.R_f$$

Therefore, the closed-loop gain of the circuit, A_{CL}, is

$$A_{CL} = V_{out}/V_{in} = -(I.R_f/I.R_1) = -R_f/R_1$$

The input voltage V_{in} is thus amplified by $-R_f/R_1$ times at the output. It can be noted that if the resistances of both the resistors, R_1 and R_f, are equal then the output voltage is,

$$V_{out} = -V_{in}$$

Such a circuit is known as an inverting buffer, or simply an Inverter.

Op-amp Non-Inverting Amplifier

A non-inverting amplifier is an op-amp circuit configuration which produces an amplified output signal that is in-phase with the input signal applied. A non-inverting amplifier uses a negative feedback connection, but instead of feeding the entire output signal to the input, only a part of the output signal voltage is fed back as input to the inverting input terminal of the op-amp.

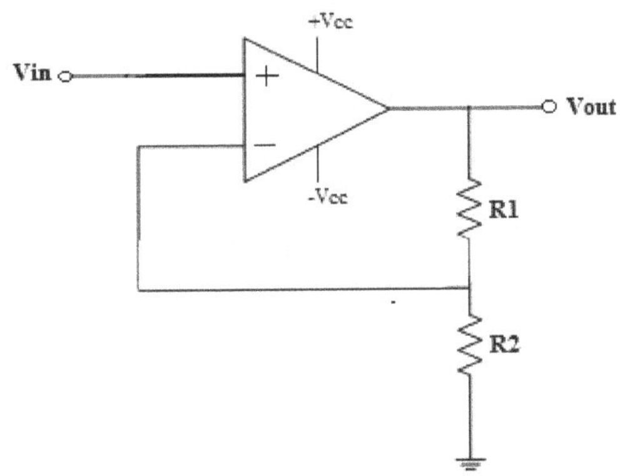

Figure: Op-amp Non-Inverting Amplifier

Above figure shows a typical non-inverting amplifier. The input signal is applied to the non-inverting input terminal and the output is fed to the inverting input terminal through a resistive potential divider network.

When a positive-going input signal is applied to the non-inverting input terminal, the output voltage will shift to keep the inverting input terminal equal to that of the input voltage applied. Hence, there will be a feedback voltage developed across resistor R_2,

$$V_{R2} = V_{IN} = I_2R_2$$

Where, I_2 is the current flowing at the junction of resistors R_1 and R_2.

$$V_{OUT} = I_2 (R_1 + R_2)$$

From the above equations of V_{IN} and V_{OUT}, the closed-loop voltage gain of the non-inverting amplifier can be calculated as,

$$A_{CL} = V_{OUT}/V_{IN}$$

$$= I_2 (R_1 + R_2) / I_2R_2$$

$$= (R_1 + R_2) / R_2$$

$$A_{CL} = 1 + (R_1/R_2)$$

The above gain equation is positive, indicating that the output will be in-phase with the applied input signal. The closed-loop voltage gain of a non-inverting amplifier is determined by the ratio of the resistors R_1 and R_2 used in the circuit.

Practical non-inverting amplifiers will have a resistor in series with the input voltage source, to keep the input current equal in both input terminals.

Instrumentation Amplifier

An instrumentation amplifier allows an engineer to adjust the gain of an amplifier circuit without having to change more than one resistor value. Compare this to the differential amplifier, which we covered previously, which requires the adjustment of multiple resistor values.

The so-called instrumentation amplifier builds on the last version of the differential amplifier to give us that capability:

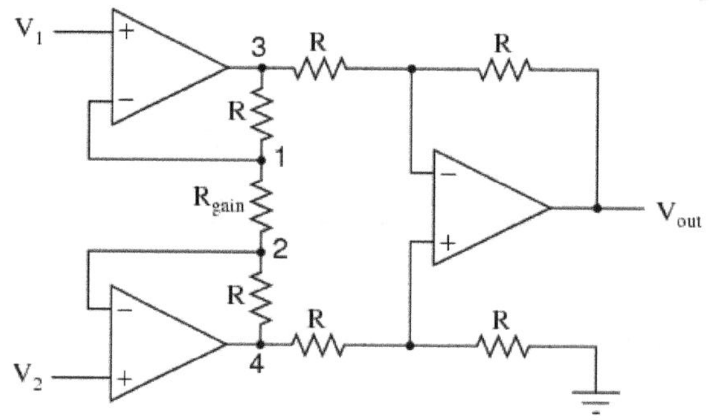

Understanding the Instrumentation Amplifier Circuit

This intimidating circuit is constructed from a buffered differential amplifier stage with three new resistors linking the two buffer circuits together. Consider all resistors to be of equal value except for R_{gain}.

The negative feedback of the upper-left op-amp causes the voltage at point 1 (top of R_{gain}) to be equal to V_1. Likewise, the voltage at point 2 (bottom of R_{gain}) is held to a value equal to V_2. This establishes a voltage drop across R_{gain} equal to the voltage difference between V_1 and V_2. That voltage drop causes a current through R_{gain}, and since the feedback loops of the two input op-amps draw no current, that same amount of current through R_{gain} must be going through the two "R" resistors above and below it.

This produces a voltage drop between points 3 and 4 equal to:

$$V_{3.4} = (V_2 - V_1)\left(1 + \frac{2R}{R_{gain}}\right)$$

The regular differential amplifier on the right-hand side of the circuit then takes this voltage drop between points 3 and 4 and amplifies it by a gain of 1 (assuming again that all "R" resistors are of equal value).

Advantages of the Instrumentation Amplifier

Though this looks like a cumbersome way to build a differential amplifier, it has the distinct advantages of possessing extremely high input impedances on the V_1 and V_2 inputs (because they connect straight into the noninverting inputs of their respective op-amps), and adjustable gain that can be set by a single resistor.

Manipulating the above formula a bit, we have a general expression for overall voltage gain in the instrumentation amplifier:

$$A_V = \left(1 + \frac{2R}{R_{gain}}\right)$$

Though it may not be obvious by looking at the schematic, we can change the differential gain of the instrumentation amplifier simply by changing the value of one resistor: R_{gain}.

We could still change the overall gain by changing the values of some of the other resistors, but this would necessitate *balanced* resistor value changes for the circuit to remain symmetrical. Please note that the lowest gain possible with the above circuit is obtained with R_{gain} completely open (infinite resistance), and that gain value is 1.

Analog Multiplier

An analog multiplier is a electronic device that produces an output voltage or current that is proportional to the product of two or more independent input voltages or currents. In addition to multiplying and dividing, multipliers can perform other mathematical functions: squaring, square-rooting, logarithm and modulation functions. The most applications include processes where a realtime response is required. Analog multiplier circuit symbol is showed in Fig.

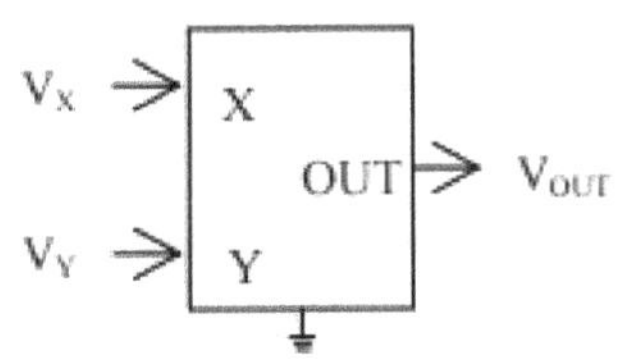

Figure: Circuit symbol of analog multiplier

The general form of the ideal multiplier transfer function is:

$$V_{out} = \frac{1}{k}.V_x.V_y$$

where, V_{out} is output voltage in volts, V_X is input voltage applied to X input, V_Y is input voltage applied to Y input, k is constant (usually 10V). If constant k is 10 then $V_{out} = \frac{1}{k}.V_x.V_y$ becomes:

$$V_{out} = \frac{V_x.V_y}{10}$$

Multipliers are always designed to have a dimensional constant k in the value of 10, so either input can have any value in the 10V range without causing the output to exceed 10V. For applications in which the maximum range of the inputs is less than 10V it is better to use smaller value of k. Many multipliers have an option to adjust dimensional constant with external components connected to appropriate pins of integrated circuit.

Classification of analog multipliers: Analog multipliers are classified according to the number of quadrants in which they can operate. The quadrants are the four quadrants of the Cartesian coordinate system as you can see in figure.

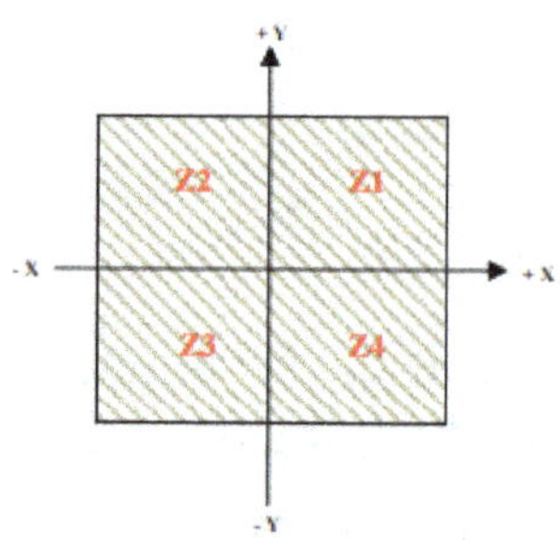

Figure: Operational quadrants of analog multiplier

Classification:

1. One-quadrant multipliers - all input signals and output signal have the same polarity (Z1 in figure).
2. Two-quadrant multipliers - one of input signals must be one-polarity and the other can be of both polarities. Output signal can be either positive or negative (Z1 and Z2 in figure).
3. Four-quadrant multipliers - allow operation with all input and output signal polarities.

Digital Electronics

Digital electronics are those electronics systems that use a digital signal instead of an analog signal. Digital electronics are the most common representation of Boolean algebra and are the basis of all digital circuits for computers, mobile phones, and numerous other consumer products.

The most common fundamental unit of digital electronics is the logic gate. By combining numerous logic gates (from tens to hundreds of thousands) more complex systems can be created. The complex system of digital electronics is collectively referred to as a digital circuit.

To most electronic engineers, the terms "digital circuit", "digital system" and "logic" are interchangeable in the context of digital circuits.

Advantages

The usual advantages of digital circuits when compared to analog circuits are:

- Digital systems interface well with computers and are easy to control with software. It is often possible to add new features to a digital system without changing hardware, and to do this remotely, just by uploading new software. Design errors or bugs can be worked-around with a software upgrade, after the product is in customer hands.
- Information storage can be much easier in digital systems than in analog ones. In particular, the great noise-immunity of digital systems makes it possible to store data and retrieve it later without degradation. In an analog system, aging and wear and tear will degrade the information in storage, but in a digital system, as long as the wear and tear is below a certain level, the information can be recovered perfectly.

Robustness

One of the primary advantages of digital electronics stems from it simply being digital: robustness. The robustness stems from the fact that if the noise is less than the noise margin then the system performs as if there were no noise at all. This is a necessary and sufficient property for a circuit to be considered digital. However, if the noise exceeds this level, the circuit can give unexpected and undesired results — perhaps catastrophic incorrect results.

Digital signals can thus be regenerated to achieve lossless data transmission, within certain limits. Analog signal transmission and processing, by contrast, always introduces noise.

Theoretically, there is no data-loss when copying digital data. This is a great advantage over analog systems, which faithfully reproduce every bit of noise that makes its way into the signal.

Disadvantages

Digital circuits use more energy than analog circuits to accomplish the same calculations and signal processing tasks, thus producing more heat as well. In portable or battery-powered systems this can be a major limiting factor, but in a situation where power is plentiful, a digital system is often preferred because of all the advantages listed above, especially that of (re-)programmability and ease of upgrading without requiring hardware changes.

A particular example is the cellular telephone, which being a battery-powered portable device, uses a low-power analog front-end to acquire and tune in the radio signal from the base station. The base station, being in a fixed location with access to the power grid, can afford to use power-hungry software-defined (digital) radio techniques that digitize the signal essentially at the antenna (after wideband filtering and downconversion to intermediate frequency) and performs all channelization and demodulation via software-driven calculations. Such base stations can be reprogrammed, potentially via remote control, to process the signals used in future cellular standards as those standards become available.

Digital circuits are sometimes more expensive, especially in small quantities.

Fragility

Digital systems can be fragile, in that if a single piece of digital data is lost or misinterpreted, the meaning of large blocks of related data can completely change. This problem can be mitigated by designing the digital system for robustness. For example, a parity bit or other error-detecting or error-correcting code can be inserted into the signal path so that if less than a certain fraction of the data is corrupted, the system can determine that this has occurred and possibly uncorrupt the data, or ask for the corrupted data to be resent. In a state-machine, the state transition logic can be designed to catch all unused states and trigger a reset sequence or other error recovery routine. For example, it is standard practice in embedded software design to fill unused program memory with interrupt instructions that point to an error recovery routine, to help guard against a failure that corrupts the microcontroller's instruction pointer which could otherwise cause random code to be executed.

Analog Issues in Digital Circuits

Digital circuits are made from analog components, and care has to be taken in design so that the analog nature of these underlying components don't dominate over the desired digital behavior. In particular, attention must be paid to all noise and timing margins, to parasitic inductances and capacitances, to proper filtering of power and ground connections, to electromagnetic coupling amongst datalines, and many other details. Inattention to these can cause intermittent problems such as "glitches", vanishingly-fast pulses that may trigger some logic but not others, "runt pulses" that do not reach valid switching (threshold) voltages, or unexpected ("undecoded") combinations of logic states.

A corollary of the fact that digital circuits are made from analog components is the fact that digital circuits are slower to perform calculations than analog circuits that occupy a similar amount of physical space and consume the same amount of power. However, the digital circuit will perform the calculation with much better repeatability, due to the high noise immunity of digital circuitry.

Construction

A digital circuit is often constructed from small electronic circuits called logic gates. Each logic gate represents a function of boolean logic. A logic gate is an arrangement of electrically controlled switches. The output is an electrical flow or voltage, that can, in turn, control more logic gates. Logic gates often use the fewest number of transistors in order to reduce their size, power consumption and cost, and increase their reliability. Manufactured as integrated circuits, they are the least expensive implementation when made in large volumes. They are usually designed by engineers using electronic design automation software.

Another form of digital circuit is constructed from lookup tables, (many sold as "programmable logic devices", though other kinds of PLDs exist). Lookup tables can perform all the same functions as machines based on logic gates, but lookup tables can be easily reprogrammed without changing the wiring. This means that a designer can often repair errors without changing the arrangement of wires. Therefore, in small volume products, programmable logic devices are often the preferred solution. They are usually designed by engineers using electronic design automation software.

When the volumes are medium to large, and the logic can be slow, or involves complex algorithms or sequences, often a small microcontroller is programmed to make an embedded system. These are usually programmed by software engineers.

When only one digital circuit is needed, and its design is totally customized, as for a factory production line controller, the conventional solution is a programmable logic controller, or PLC. These are usually programmed by electricians, using ladder logic.

Structure of Digital Systems

Engineers use many methods to minimize logic functions, in order to reduce the complexity, and thus the number of errors and the expense of digital circuits. The most widely used method is the application of a minimization algorithm within a CAD system, although historically, truth tables, Karnaugh Maps, and Boolean algebra have also been used.

Representations are crucial to an engineer's design of digital circuits. Some analysis methods only work with particular representations.

The classical way to represent a digital circuit is with an equivalent set of logic gates. Another way, often with the least electronics, is to construct an equivalent system of electronic switches (usually transistors). One of the easiest ways is to simply have a memory containing a truth table. The inputs are fed into the address of the memory, and the data outputs of the memory become the outputs.

For automated analysis, these representations have digital file formats that can be processed by computer programs. Most digital engineers are very careful to select computer programs ("tools") with compatible file formats.

To choose representations, engineers consider types of digital systems. Most digital systems divide into "combinatorial systems" and "sequential systems". A combinatorial system always presents the same output when given the same inputs. It is basically a representation of a set of logic functions.

A sequential system is a combinatorial system with some of the outputs fed back as inputs. This makes the digital machine perform a "sequence" of operations. The simplest sequential system is probably a flip flop, a mechanism that represents a binary digit or "bit".

Sequential systems are often designed as state machines. In this way, engineers can design a system's gross behavior, and even test it in a simulation, without considering all the details of the logic functions.

Sequential systems divide into two further subcategories. "Synchronous" sequential systems change state all at once, when a "clock" signal changes state. "Asynchronous" sequential systems propagate changes whenever inputs change. Synchronous sequential systems are made of well-characterized asynchronous circuits such as flip-flops, that change only when the clock changes, and which have carefully designed timing margins.

The usual way to implement a synchronous sequential state machine is divide it into a piece of combinatorial logic and a set of flip flops called a "state register." Each time a clock signal ticks, the state register captures the feedback generated from the previous state of the combinatorial logic, and feeds it back as an unchanging input to the combinatorial part of the state machine. The fastest rate of the clock is set by the most time-consuming logic calculation in the combinatorial logic.

The state register is just a representation of a binary number. If the states in the state machine are numbered (easy to arrange), the logic function is just some logic that produces the number of the next state.

In comparison, asynchronous systems are very hard to design because all possible states, in all possible timings must be considered. The usual method is to construct a table of the minimum and maximum time that each such state can exist, and then adjust the circuit to minimize the number of such states, and force the circuit to periodically wait for all of its parts to enter a compatible state. (This is called "self-resynchronization.") Without such careful design, it is easy to accidentally produce asynchronous logic that is "unstable", that is, real electronics will have unpredictable results because of the cumulative delays caused by small variations in the values of the electronic components. Certain circuits (such as the synchronizer flip-flops, switch debouncers, and the like which allow external unsynchronized signals to enter synchronous logic circuits) are inherently asynchronous in their design and must be analyzed as such.

More generally, many digital systems are data flow machines. These are usually designed using synchronous register transfer logic, using hardware description languages such as VHDL or Verilog.

In register transfer logic, binary numbers are stored in groups of flip flops called registers. The outputs of each register are a bundle of wires called a "bus" that carries that number to other calculations. A calculation is simply a piece of combinatorial logic. Each calculation also has an output bus, and these may be connected to the inputs of several registers. Sometimes a register will have a multiplexer on its input, so that it can store a number from any one of several buses. Alternatively, the outputs of several items may be connected to a bus through buffers that can turn off the output of all of the devices except one. A sequential state machine controls when each register accepts new data from its input.

The most general-purpose register-transfer logic machine is a computer. This is basically an automatic binary abacus. The control unit of a computer is usually designed as a microprogram run by a microsequencer. A microprogram is much like a player-piano roll. Each table entry or "word" of the microprogram commands the state of every bit that controls the computer. The sequencer then counts, and the count addresses the memory or combinatorial logic machine that contains the microprogram. The bits from the microprogram control the arithmetic logic unit, memory and other parts of the computer, including the microsequencer itself.

In this way, the complex task of designing the controls of a computer is reduced to a simpler task of programming a relatively independent collection of much simpler logic machines.

Computer architecture is a specialized engineering activity that tries to arrange the registers, calculation logic, buses and other parts of the computer in the best way for some purpose. Computer architects have applied large amounts of ingenuity to computer design to reduce the cost and increase the speed and immunity to programming errors of computers. An increasingly common goal is to reduce the power used in a battery-powered computer system, such as a cell-phone. Many computer architects serve an extended apprenticeship as microprogrammers.

"Specialized computers" are usually a conventional computer with a special-purpose microprogram.

Automated Design Tools

To save costly engineering effort, much of the effort of designing large logic machines has been automated. The computer programs are called "electronic design automation tools.

Simple truth table-style descriptions of logic are often optimized with EDA that automatically produces reduced systems of logic gates or smaller lookup tables that still produce the desired outputs.

Most practical algorithms for optimizing large logic systems use algebraic manipulations or binary decision diagrams, and there are promising experiments with genetic algorithms and annealing optimizations.

To automate costly engineering effort, some EDA can take state tables that describe state machines and automatically produce a truth table for the combinatorial part of a state machine. The state table is a piece of text that lists each state, and the conditions that can exit that state.

It is common for the truth tables of such computer-generated state-machines to be optimized with logic-minimization software. This is a simple example of how complex logic machines are broken into simpler parts. Often, real logic systems are designed as a series of sub-projects, which are combined using a "tool flow". The tool flow is usually a "script", a simplified computer language that can invoke the software design tools in the right order.

Tool flows for large logic systems such as microprocessors can be thousands of commands long, and combine the work of hundreds of engineers.

Writing and debugging tool flows is an established engineering specialty in companies that produce complex logic machines. The tool flow usually terminates in a detailed computer file or set of files that describe how to physically construct the logic machine. Often it consists of instructions to draw the transistors and wires on an integrated circuit or a printed circuit board.

Parts of tool flows are "debugged" by testing the outputs of simulated logic machines against expected inputs. The test tools take computer files with sets of inputs and outputs, and highlight discrepancies between the simulated behavior and the expected behavior.

These test data are usually called "test vectors." Often, the test vectors are preserved and used in the factory to test that newly constructed logic machines work correctly.

Design for Testability

A large logic machine (say, with more than a hundred logical variables) can have an astronomical number of possible states. Obviously, in the factory, testing every state is impractical if testing each state takes a microsecond, and there are more states than the number of microseconds since the universe began.

Fortunately, large logic machines are almost always designed as assemblies of smaller logic machines. To save time, the smaller sub-machines are isolated by permanently-installed "design for test" circuitry, and are tested independently.

One common test scheme known as "scan design" moves test bits serially (one after another) from external test equipment through one or more serial shift registers known as "scan chains". Serial scans have only one or two wires to carry the data, and minimize the physical size and expense of the infrequently-used test logic.

After all the test data bits are in place, the design is reconfigured to be in "normal mode" and one or more clock pulses are applied, to test for faults (e.g. stuck-at low or stuck-at high) and capture the test result into flip-flops and/or latches in the scan shift register(s). Finally, the result of the test is shifted out to the block boundary and compared against the predicted "good machine" result.

Another common testing scheme provides a test mode that forces some part of the logic machine to enter a "test cycle." The test cycle usually exercises large independent parts of the machine.

Waveforms in Digital Systems

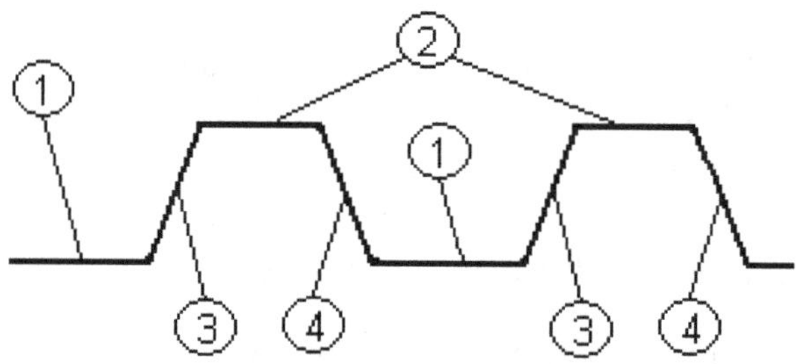

Digital signal: 1) Low level, 2) High level, 3) Rising edge, and 4) Falling edge

In computer architecture and other digital systems, a waveform that switches between two voltage levels representing the two states of a Boolean value (0 and 1) is referred to as a digital signal, even though it is an analog voltage waveform, since it is interpreted in terms of only two levels.

The clock signal is a special digital signal that is used to synchronize digital circuits. The image shown can be considered the waveform of a clock signal. Logic changes are triggered either by the rising edge or the falling edge.

Logic Gates

A Digital Logic Gate is an electronic circuit which makes logical decisions based on the combination of digital signals present on its inputs.

Digital logic gates can have more than one input, for example, inputs A, B, C, D etc., but generally only have one digital output, (Q). Individual logic gates can be connected or cascaded together to form a logic gate function with any desired number of inputs, or to form combinational and sequential type circuits, or to produce differnt logic gate functions from standard gates.

Standard commercially available digital logic gates are available in two basic families or forms, TTL which stands for *Transistor-Transistor Logic* such as the 7400 series, and CMOSwhich stands for *Complementary Metal-Oxide-Silicon* which is the 4000 series of chips. This notation of TTL or CMOS refers to the logic technology used to manufacture the integrated circuit, (IC) or a "chip" as it is more commonly called.

Digital Logic Gate

Generally speaking, TTL logic IC's use NPN and PNP type Bipolar Junction Transistors while CMOS logic IC's use complementary MOSFET or JFET type Field Effect Transistors for both their input and output circuitry.

As well as TTL and CMOS technology, simple digital logic gates can also be made by connecting together diodes, transistors and resistors to produce RTL, Resistor-Transistor logic gates, DTL, Diode-Transistor logic gates or ECL, Emitter-Coupled logic gates but these are less common now compared to the popular CMOS family.

Integrated Circuits or IC's as they are more commonly called, can be grouped together into families according to the number of transistors or "gates" that they contain. For example, a simple AND gate my contain only a few individual transistors, were as a more complex microprocessor may contain many thousands of individual transistor gates. Integrated circuits are categorised according to the number of logic gates or the complexity of the circuits within a single chip with the general classification for the number of individual gates given as:

Classification of Integrated Circuits

- Small Scale Integration or (SSI) – Contain up to 10 transistors or a few gates within a single package such as AND, OR, NOT gates.
- Medium Scale Integration or (MSI) – between 10 and 100 transistors or tens of gates within a single package and perform digital operations such as adders, decoders, counters, flip-flops and multiplexers.
- Large Scale Integration or (LSI) – between 100 and 1,000 transistors or hundreds of gates and perform specific digital operations such as I/O chips, memory, arithmetic and logic units.
- Very-Large Scale Integration or (VLSI) – between 1,000 and 10,000 transistors or thousands of gates and perform computational operations such as processors, large memory arrays and programmable logic devices.

- Super-Large Scale Integration or (SLSI) – between 10,000 and 100,000 transistors within a single package and perform computational operations such as microprocessor chips, micro-controllers, basic PICs and calculators.
- Ultra-Large Scale Integration or (ULSI) – more than 1 million transistors – the big boys that are used in computers CPUs, GPUs, video processors, micro-controllers, FPGAs and complex PICs.

While the "ultra large scale" ULSI classification is less well used, another level of integration which represents the complexity of the Integrated Circuit is known as the *System-on-Chip* or (SOC) for short. Here the individual components such as the microprocessor, memory, peripherals, I/O logic etc, are all produced on a single piece of silicon and which represents a whole electronic system within one single chip, literally putting the word "integrated" into integrated circuit.

These complete integrated chips which can contain up to 100 million individual silicon-CMOS transistor gates within one single package are generally used in mobile phones, digital cameras, micro-controllers, PIC's and robotic type applications.

Digital Logic States

The Digital Logic Gate is the basic building block from which all digital electronic circuits and microprocessor based systems are constructed from. Basic digital logic gates perform logical operations of AND, OR and NOT on binary numbers.

In digital logic design only two voltage levels or states are allowed and these states are generally referred to as Logic "1" and Logic "0", or HIGH and LOW, or TRUE and FALSE. These two states are represented in Boolean Algebra and standard truth tables by the binary digits of "1" and "0" respectively.

A good example of a digital state is a simple light switch. The switch can be either "ON" or "OFF", one state or the other, but not both at the same time. Then we can summarise the relationship between these various digital states as being:

Boolean Algebra	Boolean Logic	Voltage State
Logic "1"	TRUE (T)	HIGH (H)
Logic "0"	FALSE (F)	LOW (L)

Most digital logic gates and digital logic systems use "Positive logic", in which a logic level "0" or "LOW" is represented by a zero voltage, 0v or ground and a logic level "1" or "HIGH" is represented by a higher voltage such as +5 volts, with the switching from one voltage level to the other, from either a logic level "0" to a "1" or a "1" to a "0" being made as quickly as possible to prevent any faulty operation of the logic circuit.

There also exists a complementary "Negative Logic" system in which the values and the rules of a logic "0" and a logic "1" are reversed but in this tutorial section about digital logic gates we shall only refer to the positive logic convention as it is the most commonly used.

In standard TTL (transistor-transistor logic) IC's there is a pre-defined voltage range for the input and output voltage levels which define exactly what is a logic "1" level and what is a logic "0" level and these are shown.

TTL Input and Output Voltage Levels

There are a large variety of logic gate types in both the bipolar 7400 and the CMOS 4000 families of digital logic gates such as 74Lxx, 74LSxx, 74ALSxx, 74HCxx, 74HCTxx, 74ACTxx etc, with each one having its own distinct advantages and disadvantages compared to the other. The exact switching voltage required to produce either a logic "0" or a logic "1" depends upon the specific logic group or family.

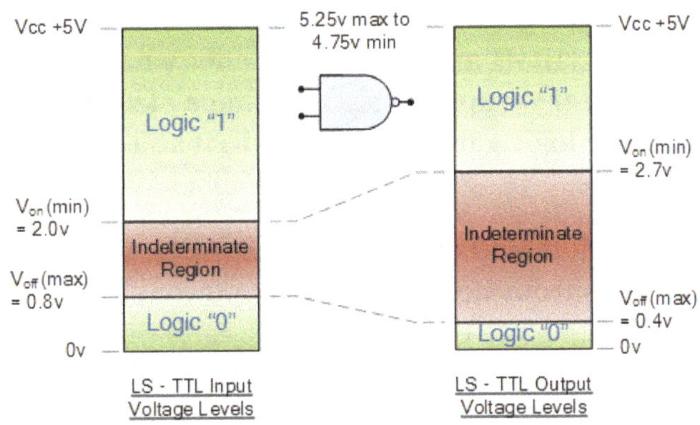

However, when using a standard +5 volt supply any TTL voltage input between 2.0v and 5v is considered to be a logic "1" or "HIGH" while any voltage input below 0.8v is recognised as a logic "0" or "LOW". The voltage region in between these two voltage levels either as an input or as an output is called the *Indeterminate Region* and operating within this region may cause the logic gate to produce a false output.

The CMOS 4000 logic family uses different levels of voltages compared to the TTL types as they are designed using field effect transistors, or FET's. In CMOS technology a logic "1" level operates between 3.0 and 18 volts and a logic "0" level is below 1.5 volts. Then the following table shows the difference between the logic levels of traditional TTL and CMOS logic gates.

TTL and CMOS Logic Levels

Device Type	Logic 0	Logic 1
TTL	0 to 0.8v	2.0 to 5v (V_{CC})
CMOS	0 to 1.5v	3.0 to 18v (V_{DD})

Then from the above observations, we can define the ideal TTL digital logic gate as one that has a "LOW" level logic "0" of 0 volts (ground) and a "HIGH" level logic "1" of +5 volts and this can be demonstrated as:

Ideal TTL Digital Logic Gate Voltage Levels

Where the opening or closing of the switch produces either a logic level "1" or a logic level "0" with the resistor R being known as a "pull-up" resistor.

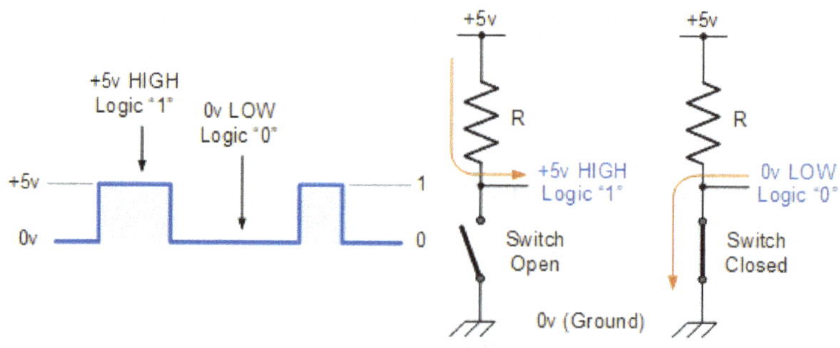

Digital Logic Noise

However, between these defined HIGH and LOW values lies what is generally called a "no-man's land" (the blue area's above) and if we apply a signal voltage of a value within this no-man's land area we do not know whether the logic gate will respond to it as a level "0" or as a level "1", and the output will become unpredictable.

Noise is the name given to a random and unwanted voltage that is induced into electronic circuits by external interference, such as from nearby switches, power supply fluctuations or from wires and other conductors that pick-up stray electromagnetic radiation. Then in order for a logic gate not to be influence by noise in must have a certain amount of noise margin or noise immunity.

Digital Logic Gate Noise Immunity

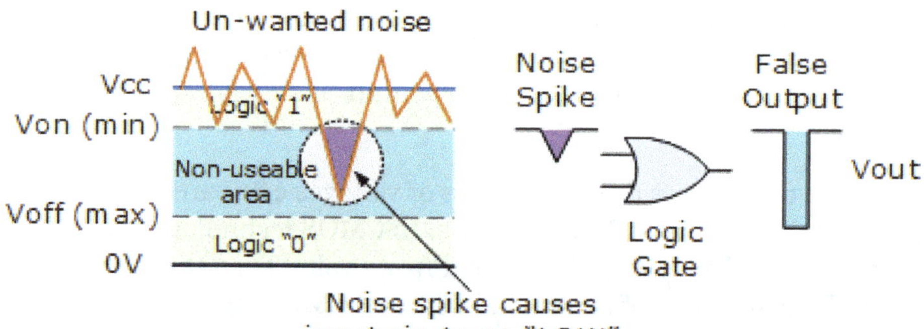

In the example above, the noise signal is superimposed onto the Vcc supply voltage and as long as it stays above the minimum level ($V_{ON(min)}$) the input an corresponding output of the logic gate are unaffected. But when the noise level becomes large enough and a noise spike causes the HIGH voltage level to drop below this minimum level, the logic gate may interpret this spike as a LOW level input and switch the output accordingly producing a false output switching. Then in order for the logic gate not to be affected by noise it must be able to tolerate a certain amount of unwanted noise on its input without changing the state of its output.

Simple Basic Digital Logic Gates

Simple digital logic gates can be made by combining transistors, diodes and resistors with a simple example of a Diode-Resistor Logic (DRL) AND gate and a Diode-Transistor Logic (DTL) NAND gate given.

Diode-Resistor Circuit	Diode-Transistor circuit
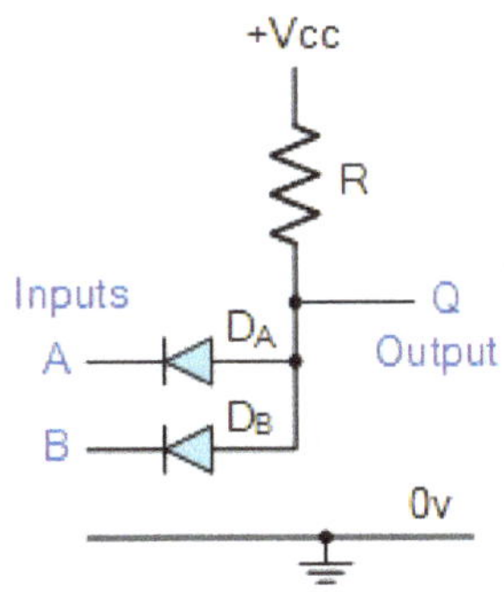 2-input AND Gate	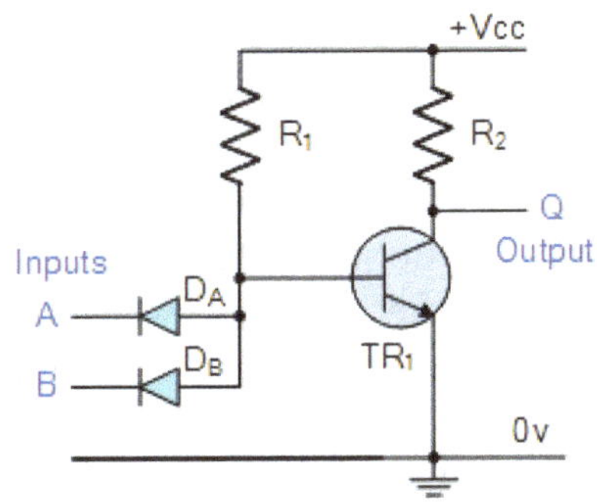 2-input NAND Gate

The simple 2-input Diode-Resistor AND gate can be converted into a NAND gate by the addition of a single transistor inverting (NOT) stage. Using discrete components such as diodes, resistors and transistors to make digital logic gate circuits are not used in practical commercially available logic IC's as these circuits suffer from propagation delay or gate delay and also power loss due to the pull-up resistors.

Another disadvantage of diode-resistor logic is that there is no "Fan-out" facility which is the ability of a single output to drive many inputs of the next stages. Also this type of design does not turn fully "OFF" as a Logic "0" produces an output voltage of 0.6v (diode voltage drop), so the following TTL and CMOS circuit designs are used instead.

Basic TTL Logic Gates

The simple Diode-Resistor AND gate above uses separate diodes for its inputs, one for each input. As a transistor is made up off two diode circuits connected together representing an NPN or a PNP device, the input diodes of the DTL circuit can be replaced by one single NPN transistor with multiple emitter inputs as shown.

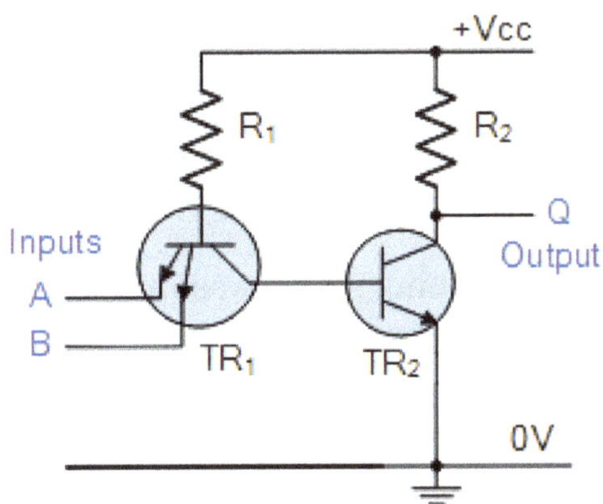

2-input NAND Gate

As the NAND gate contains a single stage inverting NPN transistor circuit (TR_2) an output logic level "1" at Q is only present when both the emitters of TR_1 are connected to logic level "0" or ground allowing base current to pass through the PN junctions of the emitter and not the collector. The multiple emitters of TR_1 are connected as inputs thus producing a NAND gate function.

In standard TTL logic gates, the transistors operate either completely in the "cut off" region, or else completely in the saturated region, Transistor as a Switch type operation.

Emitter-Coupled Digital Logic Gate

Emitter Coupled Logic or *ECL* is another type of digital logic gate that uses bipolar transistor logic where the transistors are not operated in the saturation region, as they are with the standard TTL digital logic gate. Instead the input and output circuits are push-pull connected transistors with the supply voltage negative with respect to ground.

This has the effect of increasing the speed of operation of the emitter coupled logic gates up to the Gigahertz range compared with the standard TTL types, but noise has a greater effect in ECL logic, because the unsaturated transistors operate within their active region and amplify as well as switch signals.

The "74" Sub-families of Integrated Circuits

With improvements in the circuit design to take account of propagation delays, current consumption, fan-in and fan-out requirements etc, this type of TTL bipolar transistor technology forms the basis of the prefixed "74" family of digital logic IC's, such as the "7400" Quad 2-input NAND gate, or the "7402" Quad 2-input NOR gate, etc.

Sub-families of the 74xxx series IC's are available relating to the different technologies used to fabricate the gates and they are denoted by the letters in between the 74 designation and the device number. There are a number of TTL sub-families available that provide a wide range of switching speeds and power consumption such as the 74L00 or 74ALS00 NAND gate, were the "L" stands for "Low-power TTL" and the "ALS" stands for "Advanced Low-power Schottky TTL" and these are listed below:

- 74xx or 74Nxx: Standard TTL – These devices are the original TTL family of logic gates introduced in the early 70's. They have a propagation delay of about 10ns and a power consumption of about 10mW. Supply voltage range: 4.75 to 5.25 volts.
- 74Lxx: Low Power TTL – Power consumption was improved over standard types by increasing the number of internal resistances but at the cost of a reduction in switching speed. Supply voltage range: 4.75 to 5.25 volts.
- 74Hxx: High Speed TTL – Switching speed was improved by reducing the number of internal resistances. This also increased the power consumption. Supply voltage range: 4.75 to 5.25 volts.
- 74Sxx: Schottky TTL – Schottky technology is used to improve input impedance, switching speed and power consumption (2mW) compared to the 74Lxx and 74Hxx types. Supply voltage range: 4.75 to 5.25 volts.
- 74LSxx: Low Power Schottky TTL – Same as 74Sxx types but with increased internal resistances to improve power consumption. Supply voltage range: 4.75 to 5.25 volts.
- 74ASxx: Advanced Schottky TTL – Improved design over 74Sxx Schottky types optimised

to increase switching speed at the expense of power consumption of about 22mW. Supply voltage range: 4.5 to 5.5 volts.

- 74ALSxx: Advanced Low Power Schottky TTL – Lower power consumption of about 1mW and higher switching speed of 4nS compared to 74LSxx types. Supply voltage range: 4.5 to 5.5 volts.
- 74HCxx: High Speed CMOS – CMOS technology and transistors to reduce power consumption of less than 1uA with CMOS compatible inputs. Supply voltage range: 4.5 to 5.5 volts.
- 74HCTxx: High Speed CMOS – CMOS technology and transistors to reduce power consumption of less than 1uA but has increased propagation delay of about 16nS due to the TTL compatible inputs. Supply voltage range: 4.5 to 5.5 volts.

Basic CMOS Digital Logic Gate

One of the main disadvantages with the TTL digital logic gate series is that the logic gates are based on bipolar transistor logic technology and as transistors are current operated devices, they consume large amounts of power from a fixed +5 volt power supply.

Also, TTL bipolar transistor gates have a limited operating speed when switching from an "OFF" state to an "ON" state and vice-versa called the "gate" or "propagation delay". To overcome these limitations complementary MOS called "CMOS" (Complementary Metal Oxide Semiconductor) logic gates which use "Field Effect Transistors" or FET's were developed.

As these gates use both P-channel and N-channel MOSFET's as their input device, at quiescent conditions with no switching, the power consumption of CMOS gates is almost zero, (1 to 2μA) making them ideal for use in low-power battery circuits and with switching speeds upwards of 100MHz for use in high frequency timing and computer circuits.

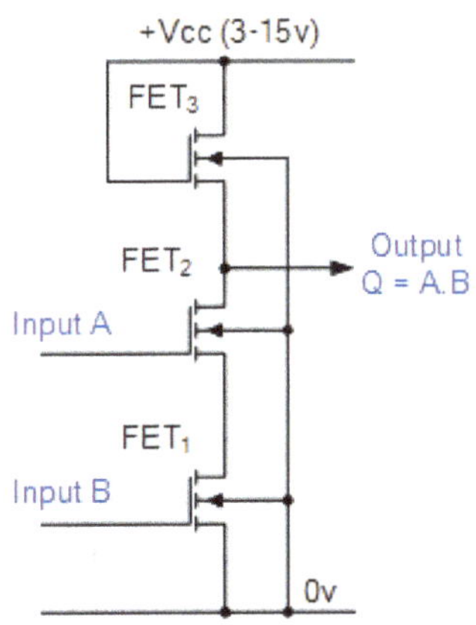

2-input NAND Gate

This CMOS gate example contains three N-channel MOSFET's, one for each input FET_1 and FET_2 and one for the output FET_3. When both the inputs A and B are at logic level "0", FET_1 and FET_2 are both switched "OFF" giving an output logic "1" from the source of FET_3.

When one or both of the inputs are at logic level "1" current flows through the corresponding FET giving an output state at Q equivalent to logic "0", thus producing a NAND gate function.

Improvements in the circuit design with regards to switching speed, low power consumption and improved propagation delays has resulted in the standard CMOS 4000 "CD" family of logic IC's being developed that complement the TTL range.

As with the standard TTL digital logic gates, all the major digital logic gates and devices are available in the CMOS package such as the CD4011, a Quad 2-input NAND gate, or the CD4001, a Quad 2-input NOR gate along with all their sub-families.

Like TTL logic, complementary MOS (CMOS) circuits take advantage of the fact that both N-channel and P-channel devices can be fabricated together on the same substrate material to form various logic functions.

One of the main disadvantage with the CMOS range of IC's compared to their equivalent TTL types is that they are easily damaged by static electricity. Also unlike TTL logic gates that operate on single +5V voltages for both their input and output levels, CMOS digital logic gates operate on a single supply voltage of between +3 and +18 volts.

Common CMOS Sub-families include:

- 4000B Series: Standard CMOS – These devices are the original Buffered CMOS family of logic gates introduced in the early 70's and operate from a supply voltage of 3.0 to 18v d.c.
- 74C Series: 5v CMOS – These devices are pin-compatible with standard 5v TTL devices as their logic switching is implemented in CMOS but with TTL-compatible inputs. They operate from a supply voltage of 3.0 to 18v d.c.

Note that CMOS logic gates and devices are static sensitive, so always take the appropriate precautions of working on antistatic mats or grounded workbenches, wearing an antistatic wristband and not removing a part from its antistatic packaging until required.

AND Gate

In digital electronics there are several logical gates which work or operate on different logical operations, say logical addition, logical multiplication etc. AND Gate is a logical gate which is widely used having two or more inputs and a single output. This gate works or operates on logical multiplication rules. In this gate if either of the inputs is low (0), then the output is also low, but if all the inputs are high (1) the output will also be high (1).

There are many integrated circuit which works on this logic we will come to it later. AND gate performs multiplication operation of binary digits. We also know there are two binary digits 1 and 0. In multiplying 0 with 0 we will get 0, 1 with 0 or 0 with 1 we will get 0. Only we get 1 when 1 is multiplied by 1.

$$0 \times 0 = 0$$

$$0 \times 1 = 0$$

$$1 \times 0 = 0$$

$$1 \times 1 = 1$$

In other words, an AND gate is a digital device which produces high output only when all inputs are high and produces low output at all other inputs conditions. High digital signal means logically 1 and low digital signal means logically 0.

An AND gate may have any number of input probes but only one output probe. A two input AND gate is logically represented as:

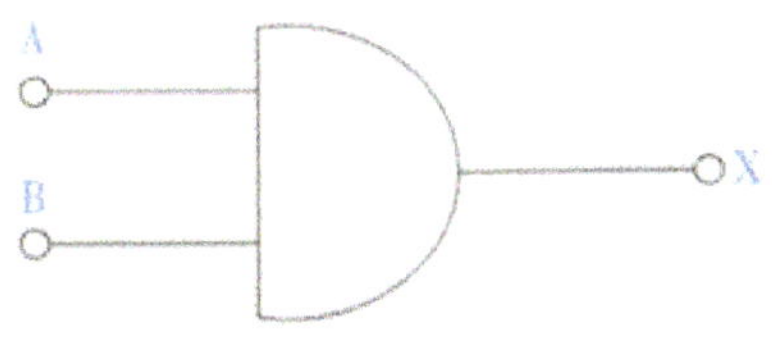

Where, A and B represent input of the gate and X represents output. A, B and X either be 1 or 0 logically.

The logical operation of AND gate hence can be represented as $AB = X$.

All multiplication combination of A and B can be represented in tabular form as follows,

A	B	C
0	0	0
0	1	0
1	0	0
1	1	1

This table is popularly known as truth table.

If instead of two there are three are numbers of inputs, the logical symbol and truth table of the gate would be,

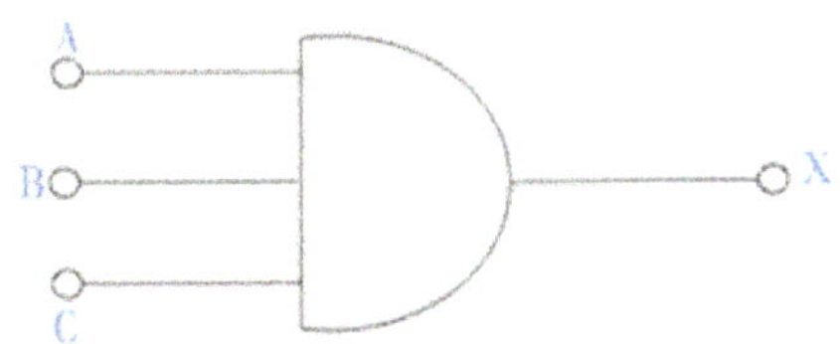

A	B	C	X
0	0	0	0
0	0	1	0
0	1	0	0
0	1	1	0
1	0	0	0
1	0	1	0
1	1	0	0
1	1	1	1

Practical Circuit of AND Gate

Diode AND Gate

Normally an AND gate is designed by either diodes or transistors. While, diodes are used to design AND gate, it is called diode AND gate. The basic circuit of a diode AND gate is shown below.

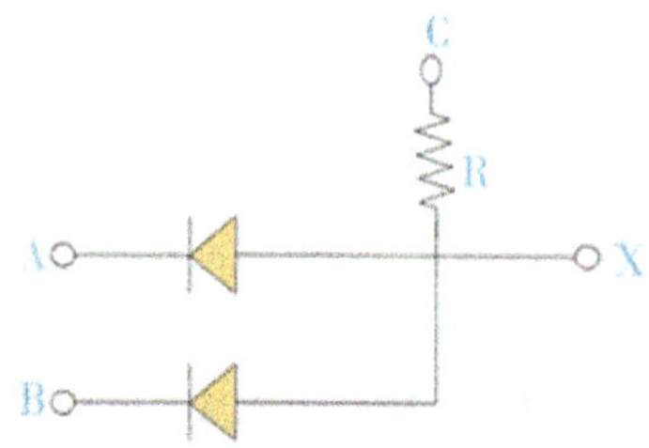

In the circuit we first apply +5V at C. Now if we apply +5V at A and B, both of the diodes are reversed biased and hence behave both diodes as OFF or open circuit.

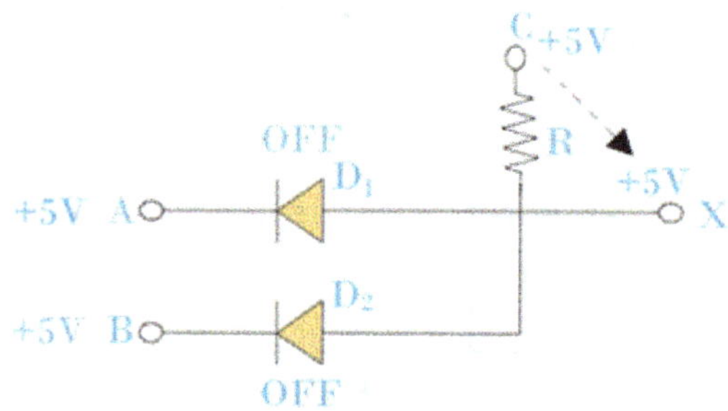

At this situation as both diodes are OFF, no current will flow through resistor R and voltage of C (+5V) will also appears at X. As the supply voltage +5V appears at X, the output of the circuit is considered as high or logical 1.

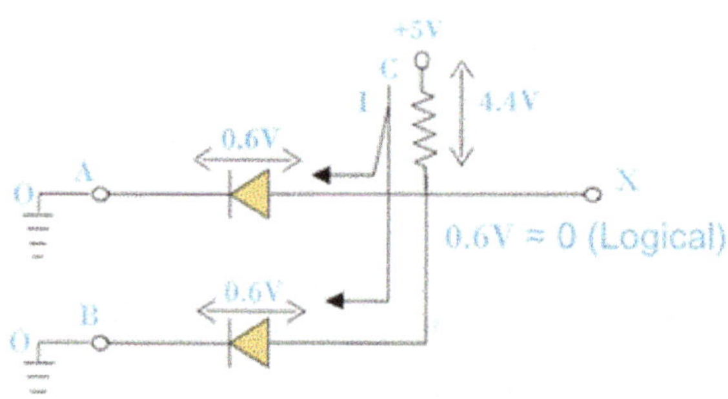

Now, if either point A or B or both are applied with 0 Volt or they are grounded, respective diode will become forward biased and hence behaves as 'ON' or short circuited. At this condition, supply voltage +5V at point C will get path through either of diodes or both to the ground potential. As the current flowing from C to ground through resistor R, entire 5V will be dropped across the resistor and hence voltage at X will become low or logically zero. The diodes at forward biased condition do not behave as ideal short circuit; some voltage drop will be there across the forward biased diodes which is equal to forward bias voltage. This voltage drop will appear at X during low output condition, so practically low output will not be 0V it is rather 0.6 to 0.7V which is ideally considered as zero.

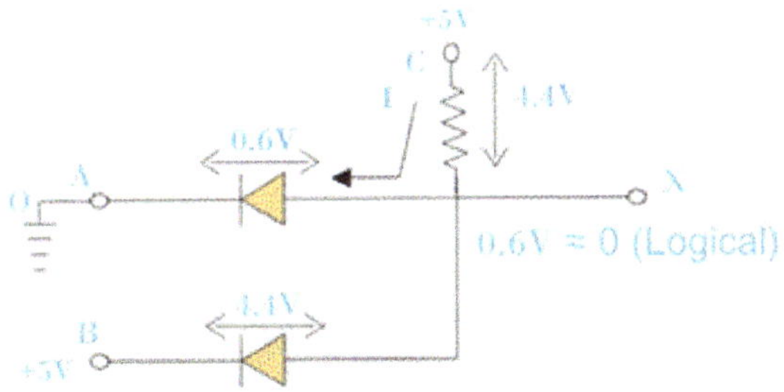

Transistor AND Gate

An AND logic gate can also be realized from transistor AND gate. The circuit diagram of transistor logic gate is shown below.

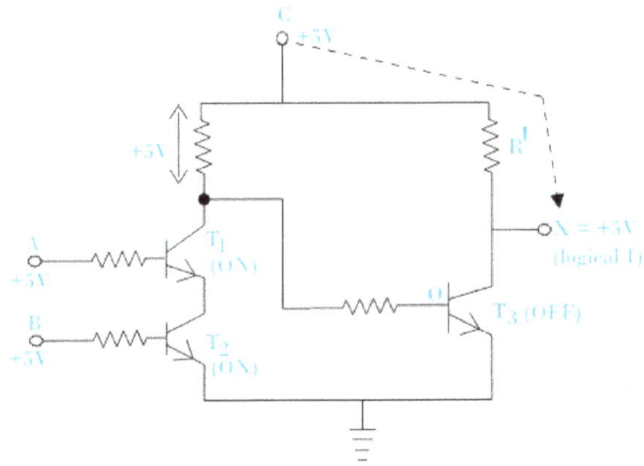

In the above circuit when A or B or both A and B are grounded or at 0V potential transistor T_1 or T_2 or both T_1 and T_2 are in OFF condition respectively. This is because terminal A and B are base terminal of transistor T_1 and T_2 respectively. Zero base voltage makes a transistor OFF. As the path through $T_1$and T_2 is open circuited base of transistor T_3 enough potential to makes T_3 ON. Current then starts flowing the supply to ground through T_3. As a result entire supply voltage will drop across R^1 and potential of terminal X will become low or logical zero.

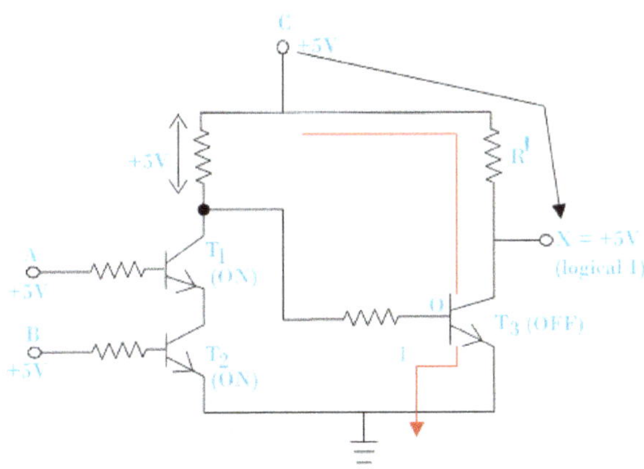

If any of the transistors T_1 and T_2 is in OFF condition, same result will come at output X as both the transistors are in series.

Now we will check what will be the logical value of X, if both A and B are at high logical value. If we apply +5V at both A and B i.e. at base of transistor T_1 and T_2 respectively.

This makes both the transistor T_1 and T_2 are in ON condition. Enter supply voltage will drop across R and the base potential of the transistor T_3 will be zero and T_3 becomes in OFF condition. As a result the supply voltage +5V appears at X and X will become logically 1 or high.

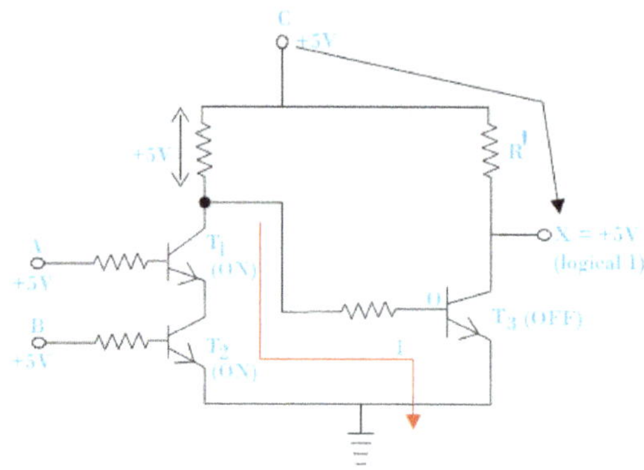

IC 7408

For AND Gate IC number in TTL is 7408. 7408 is Quad 2- input IC where four gates are present together. Let us have a look on the internal diagram of 7408. Here pin 1, 2 are the inputs of the first gate whose respective output is 3. Again 4 and 5 are the inputs of the second gate whose output is at pin 6. The inputs of fourth is pin 12 and 13 and pin 11 is its output. Pin 14 is the supply input which can be maximum 5.2 volt D.C. if input voltage be more than this it may cause damage to the IC.

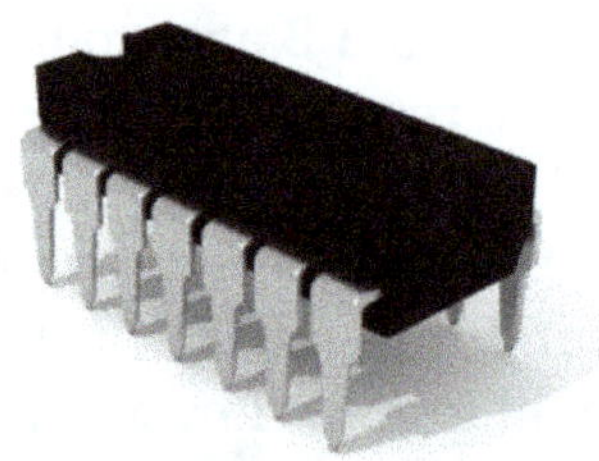

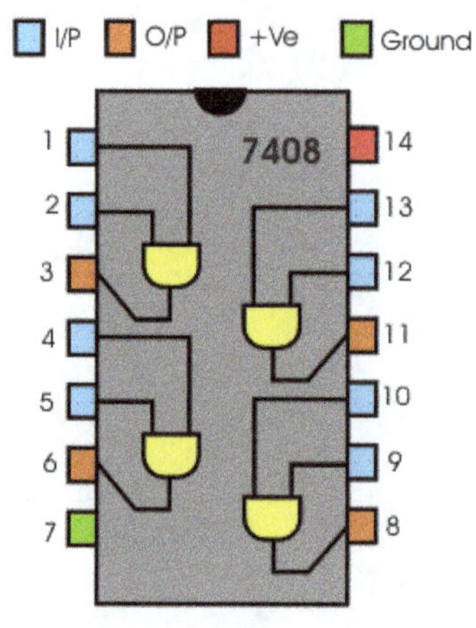

IC 4081

In CMOS logic i.e complementary MOSFET logic I.C number of AND Gate is 4081. This IC also has two inputs and one respective output. In this IC there are also 4 gates together. Now see the internal diagram of this circuit to make it more clear.

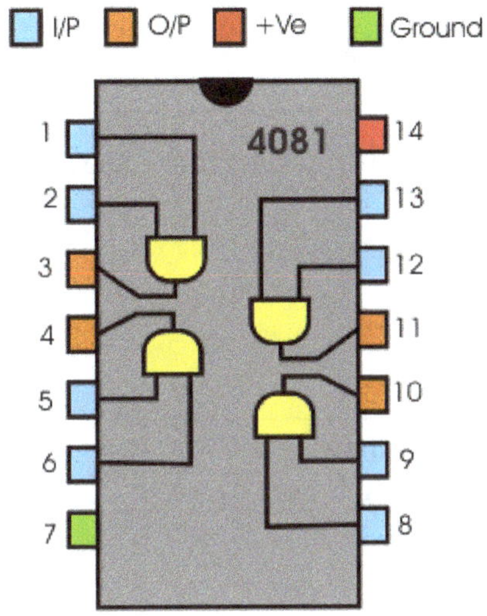

Pin 1 and 2 are the inputs of the first gate whose output is in number 3. Again Pin 5 and 6 are the inputs of the second gate whose output is at Pin 4. Pin number 7 is grounded. Pin 8 and 9 are the inputs of third gate whose output is at pin 10. Again pin 13 and 12 are the inputs of fourth AND gate whose output is at pin number 11. Pin 13 and 12 are the inputs of fourth AND Gate whose output is at pin number 11. Pin number 14 is power supply where maximum 5.2 volt D.C supply can be given to activate the IC. Here too if more voltage is given if may damage the IC. Inter circuit of CMOS and TTL differs from each other which must be noticed carefully.

OR Gate

OR gate performs logical OR operation which means outputs is logical 1 if at least one of the inputs is 1. Just like AND gate an OR gate may also have two or any more numbers of inputs but only one output. Only if all of the inputs are only in low state or logical 0 the output is low or 0 and in all other inputs conditions the output will be high or logical 1.

The logical symbol of OR gate is shown below,

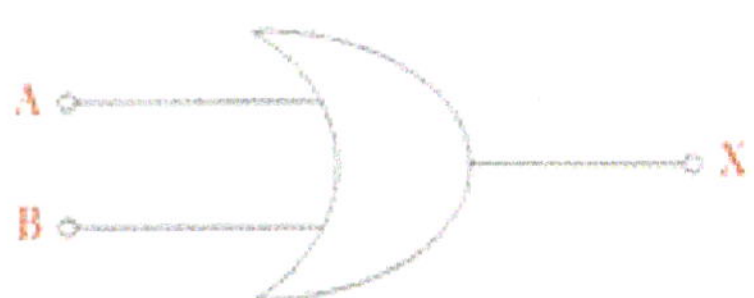

Truth Table of OR Gate

From above explanation the truth table of logical OR gate can be represented as,

Inputs		Output
A	B	X
0	0	0
0	1	1
1	0	1
1	1	1

Logical OR gate can be of more than two inputs. In that case logical symbol and truth table would be,

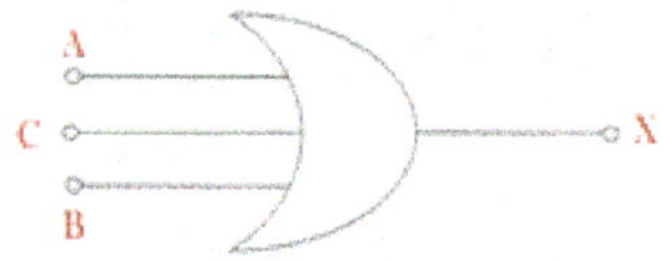

Inputs			Output
A	B	C	X
0	0	0	0
0	0	1	1
0	1	0	1
0	1	1	1
1	0	0	1
1	0	1	1
1	1	0	1
1	1	1	1

Realization of OR Gate

Diode OR Gate

A simple two inputs OR gate can be realized by using diode as follows,

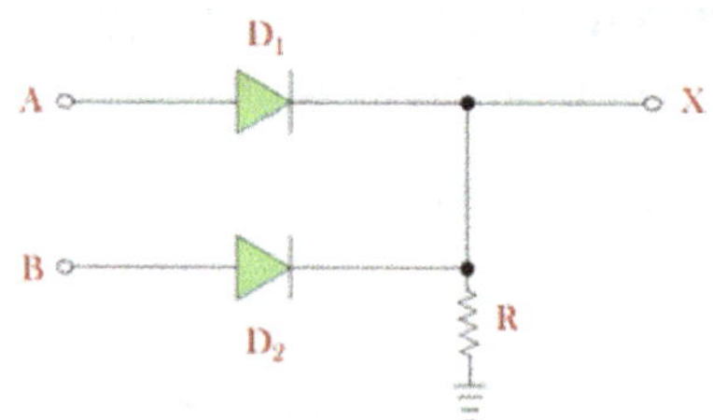

In the above circuit, if A and B are applied with 0V, there will be no voltage appears at X. When any of the inputs is given with +5V, the respective diode becomes forward biased and behaves as ideally short circuited hence this +5 V will appear at output X. +5 V means logical 1. Actually entire 5V will not appear at X, around 0.6 to 0.7 V will drop across the diode as forward bios voltage and rest of the voltage i.e. 5 – 0.6 = 4.4 V or 5 – 0.7 = 4.3 V will appear at X. This 4.4 V or 4.3 V is practically considered as logical 1.

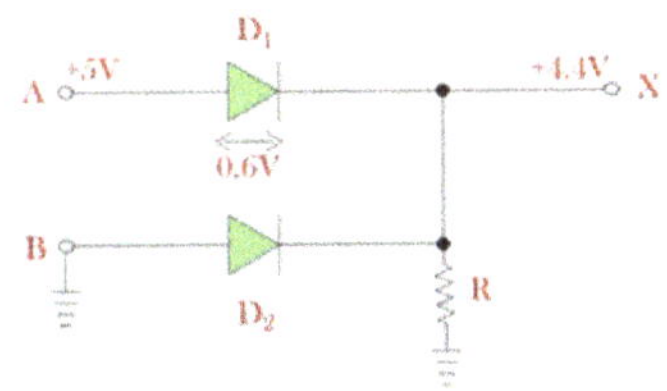

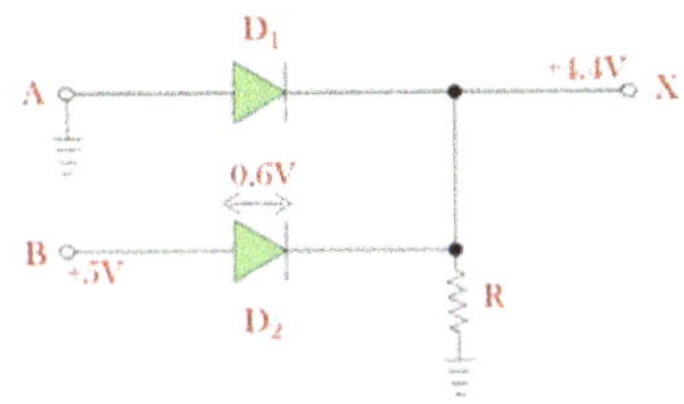

Now if both of the inputs are given with +5 V, both diodes will be forward biased. Hence, similarly 4.4 V will appear at X.

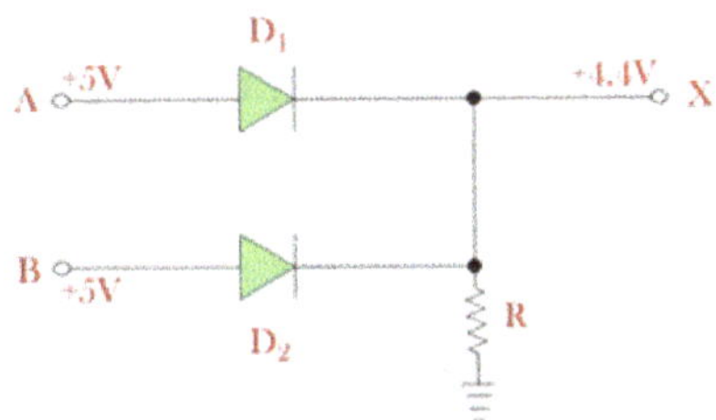

Now if both of the inputs A and B are grounded or given 0V, There will be no voltage appears at X and hence X is considered as logical 0.

Transistor OR Gate

The OR gate can also be realized by using transistor. In this case the OR gate is referred as transistor OR gate. Two inputs such OR gate is shown below,

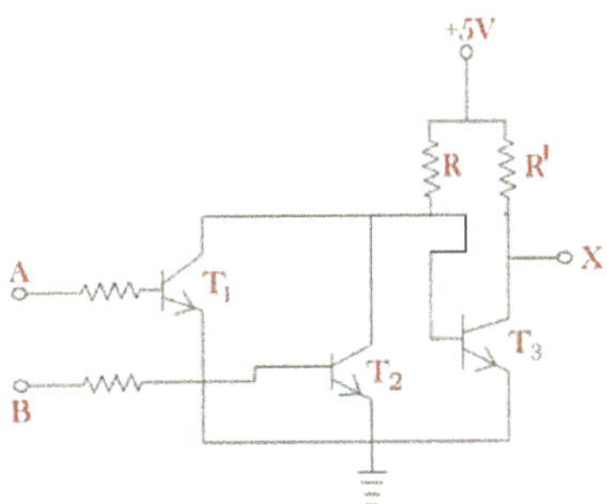

Now if A and B both are given with 0V, both of the transistor are in OFF condition, hence supply voltage + 5 V will not get path to the ground through either of the transistors, T_1 and T_2. As a result base of the transistor T_3 will get enough potential to make it ON. In that condition supply + 5 V will get path to ground through resistor R′ and transistor T_3. As the transistor T_3 is in ON condition it will behaves as ideally short circuited, hence entire supply voltage + 5 V will drop across resistor R′ and X terminal (Node) will get 0V. In practice, transistor T_3 will not be ideal short circuited it will have some voltage drop across it which will be around 0.6 – 0.7 V. This voltage will appear at node X and this 0.6 or 0.7 volt is considered as logical 0.

Now, if base terminal either of the transistors T_1 or T_2 or both are given with + 5 V, the respective transistor as both will be in ON condition. In that case supply voltage + 5 V will get path to ground through either of the transistors or both.

As a result current starts flowing to the ground from supply through this path, and entire supply voltage will drop across resistor R. So, the base of transistor T_3 will not get sufficient potential to make the transistor T_3 ON. Hence entire supply voltage will appear at X and the X becomes at high logical state or logical 1.

OR Gate ICs

OR gate is available in TTL and CMOS ICS, Where TTL is Transistor – Transistor Logic ICs and CMOS is Complementary MOSFET ICS. We will now discuss on the arrange of OR gate on both the ICs. In CMOS ICS. 4000 series is available in digital electronic 4071 is the CMOS IC number of inputs in each gate is two. 4075 is 3 – input OR gate and 4072 is 4 – input OR gate in CMOS ICs. A internal gate picture of 4071 can make you understand about this IC.

IC 4071

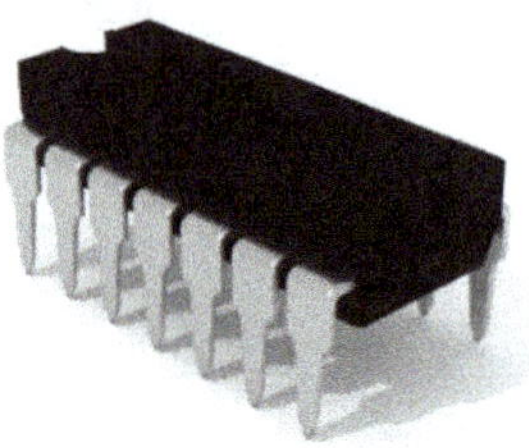

4071 is a 14 pin 1C where four or gates are fixed together having two inputs. Pin number 14 is tvcc where maximum SV DC supply is given which activates the IC. Pin number 7 is grounded. Pin 1 and 2 is the input for the first gate and 3 is the respective output for the first gate. Pin number 5 and 6 are inputs for the second gate whose output is pin number 4. For the third gate pin number 8 and 9 are the inputs whose output is pin number 10 for the last gate pin number 12 and 13 are the inputs whose respective output is at pin number 11. Thus how the IC looks internally for CMOS.

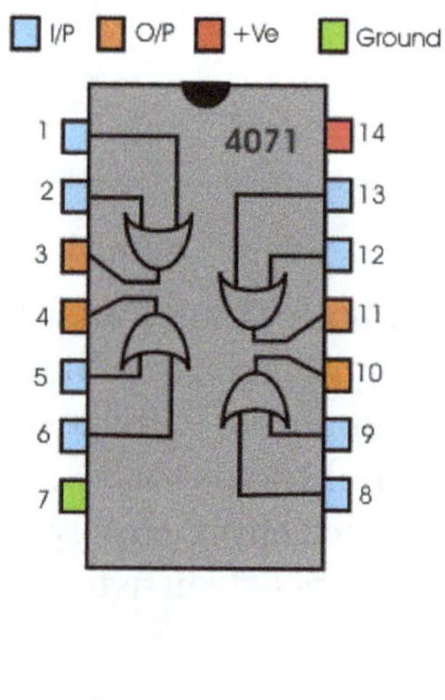

IC 7432

In TTL logic the internal arrangement differs from that incase of CMOS logics. In TTL logic 2- input OR gate IC number is 74LS32. popularly known as 7432. LS stands for low power schottky version. Again one more type of IC number is also available in market which is 74HC32 where HC stands for high speed CMOS version i.e. it has lower current consumption and wider voltage range. 74LVC32 is another low voltage CMOS version of the same. Let us have discussion on 74LS32 which is a TTL IC. In this IC there are fourteen pin. The internal gate diagram of 7432 can make your idea clear.

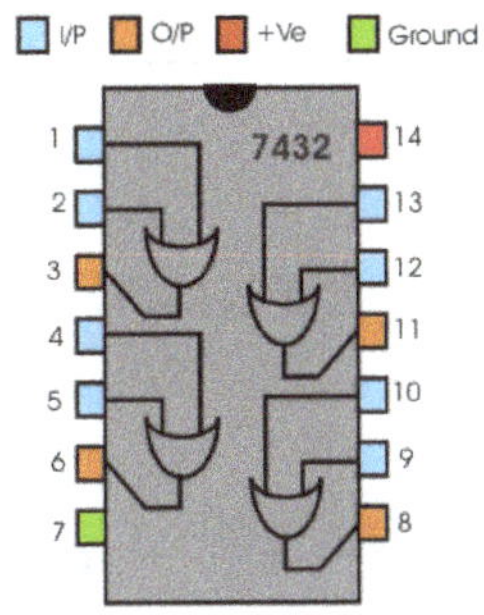

In this IC pin 1 and 2 are the inputs of first gate where the output is from pin number 3. Again pin number 4 and 5 are the inputs of second gate whose output is in pin 6. Pin 10 and 9 are the inputs of fourth gate whose output is at pn 8. The input of the last gate or fourth gate is pin 13 and 12 and pin 11 is its output. Pin 7 is ground and pin 14 is +vcc supply where again +Vcc supply where again +SVdC is the normal and maximum supply. One thing must be maintain at +SVDC. If the i/p voltage would be more than this it may cause damage to the IC.

NOT Gate

The NOT gate is a single input single output gate. This gate is also known as Inverter because it performs the inversion of the applied binary signal, i.e., it converts 0 into 1 or I into 0. In other words, the gate which has high input signal only when their input signal is low such type of gate is known as the not gate. The logic symbol for the gate is shown below.

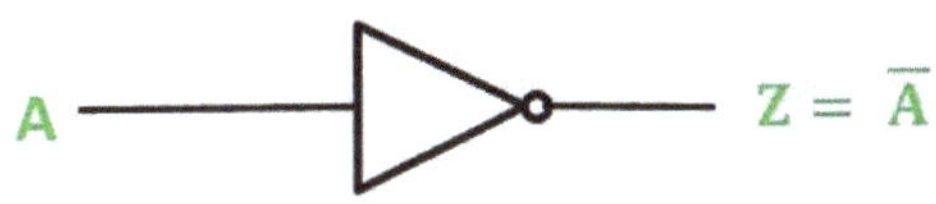

The symbol of the not gate is a triangle with a bubble on its end. This bubble is known as the inversion bubble. It gives the complement of the input signal. It is also known as the decision making devices because it has only one input.The switching circuit diagram of the NOT gate is shown below.

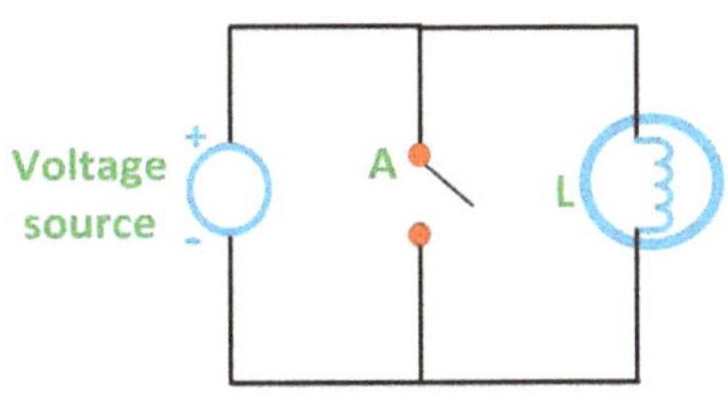

The switching circuit illustrates that the lamp will glow when switch A is open and will go off when the switch A is closed.

The truth table of the gate is given below.

A	Z
0	1
1	0

NAND Gate

A NOT gate followed by an AND gate makes a NAND gate. The basis logical construction of the NAND gate is shown below,

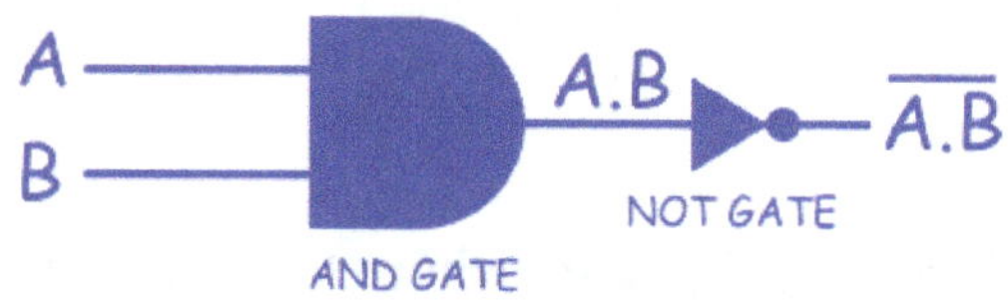

The symbol of NAND gate is similar to AND gate but one bubble is drawn at the output point of the AND gate, in the case of NAND gate. The symbol of NAND gate is shown below.

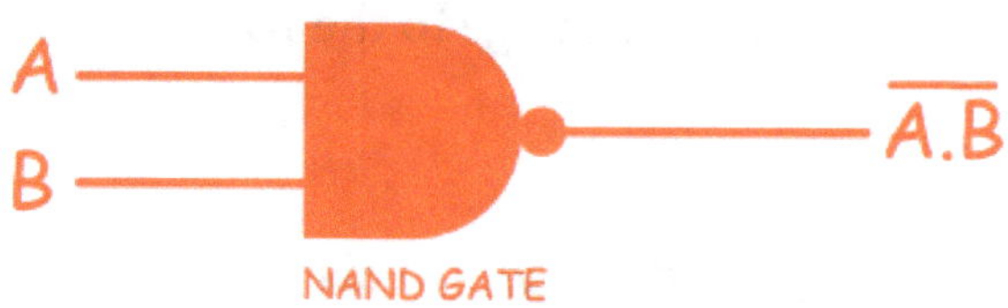

NAND gate means "not AND gate" which means, the output of this gate is just reverse of that of a similar AND gate. We know that the output of the AND gate is only high or 1 when all the inputs are high or 1. In all other cases, the output of AND gate is low or 0. In the NAND, the fact is an opposite, here, the output is only logical 0 when and only when all inputs of the gate are 1s, and in all other cases, the output of NAND gate is high or 1.

Hence, truth table of a NAND gate can be written like,

Inputs		Output
A	B	$X = \overline{\overline{A}.\overline{B}}$
0	0	1
0	1	1
1	0	1
1	1	0

Just reverse of the truth table of AND gate which is,

Inputs		Output
A	B	X=A.B
0	0	0
0	1	0
1	0	0
1	1	1

Like AND gate a NAND gate can also be more than two inputs, like 3, 4, input NAND gate.

A NAND gate is also referred as universal logic gate as all the binary operations can be realised by using only NAND gates.

There are three basic binary operations, AND, OR and NOT. By these three basic operations, one can realise all complex binary operations.

Realizing NOT Gate using NAND Gate

When both inputs of a two inputs NAND gate are zero, the output is 1, and both inputs of the NAND gate are 1, the output is 0. Hence a NOT gate can very easily be realised from NAND gates just by applying common inputs to the NAND gate or by short-circuiting all the inputs terminals of a NAND gate.

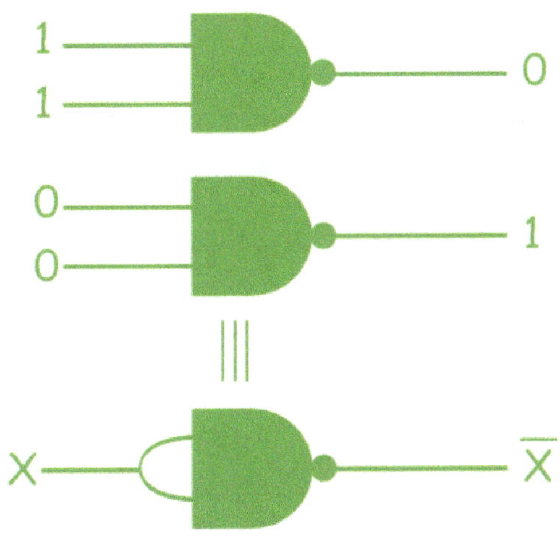

Where, X is either 1 or 0.

Realizing AND Gate using NAND Gate

A NAND gate is a NOT gate followed by an AND gate, so if we can cancel the effect of NOT gate in a NAND gate, it will become an AND gate. Hence, a NOT gate followed by a NAND gate realises an AND gate. In this case, we use the NOT gates realised from NAND gates, and we are showing the logic circuit below,

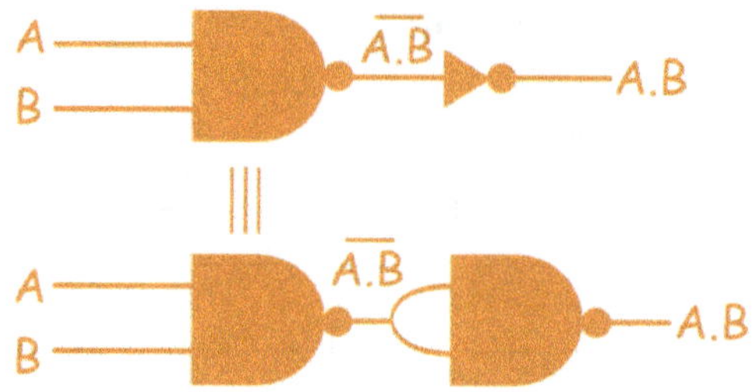

Realizing OR Gate from NAND Gate

From De Morgan Theorem we know,

$$X = \overline{\overline{A}.\overline{B}} = \overline{\overline{A}} + \overline{\overline{B}} = A + B$$

The above equation is a logical OR operation.

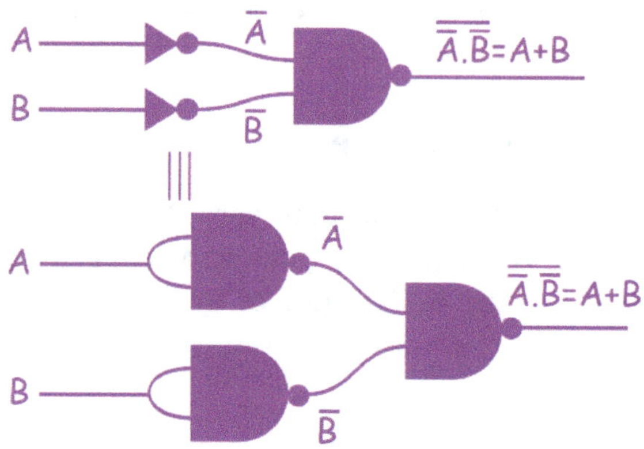

The above logic equation can be represented by gates as shown above, where inputs first inverted then passed through a third NAND gate.

The truth table of such circuit is,

Inputs		Output
A	B	$X = \overline{\overline{A}.\overline{B}}$
0	0	0
0	1	1
1	0	1
1	1	1

NOR Gate

The Logic NOR Gate gate is a combination of the digital logic OR gate and an inverter or NOT gate connected together in series.

The inclusive NOR (Not-OR) gate has an output that is normally at logic level "1" and only goes "LOW" to logic level "0" when ANY of its inputs are at logic level "1". The Logic NOR Gate is the reverse or "Complementary" form of the inclusive OR gate.

Logic NOR Gate Equivalent

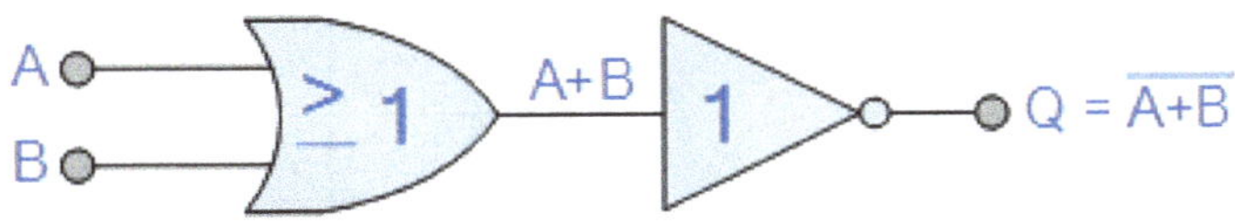

2-input "OR" gate plus a "NOT" gate

The logic or Boolean expression given for a logic NOR gate is that for *Logical Multiplication*which it performs on the *complements* of the inputs. The Boolean expression for a logic NOR gate is denoted by a plus sign, (+) with a line or *Overline*, ($\overline{\ \ }$) over the expression to signify the NOT or logical negation of the NOR gate giving us the Boolean expression of: $\overline{A+B}$ = Q.

Then we can define the operation of a 2-input digital logic NOR gate as being:

"If both A and B are NOT true, then Q is true"

Transistor NOR Gate

A simple 2-input logic NOR gate can be constructed using RTL Resistor-transistor switches connected together as shown below with the inputs connected directly to the transistor bases. Both transistors must be cut-off "OFF" for an output at Q.

Logic NOR Gates are available using digital circuits to produce the desired logical function and is given a symbol whose shape is that of a standard OR gate with a circle, sometimes called an "inversion bubble" at its output to represent the NOT gate symbol with the logical operation of the NOR gate given as.

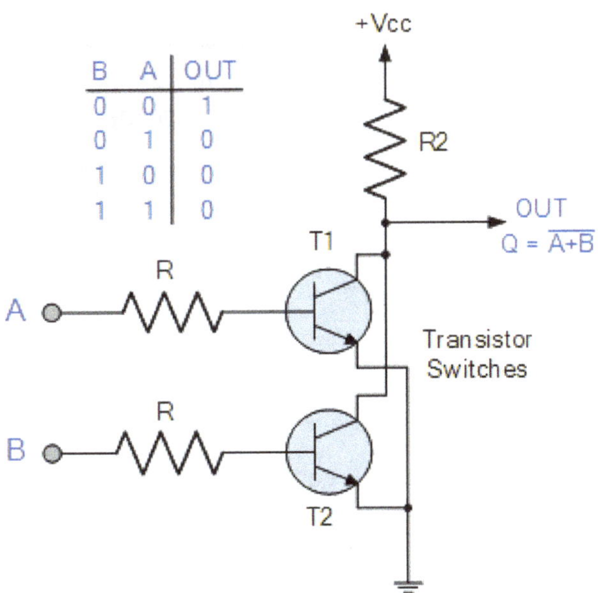

The Digital Logic "NOR" Gate

2-input NOR Gate

Symbol	Truth Table		
	B	A	Q
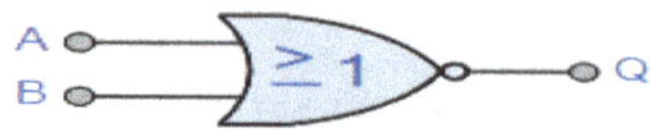	0	0	1
	0	1	0
	1	0	0
2-input NOR Gate	1	1	0
Boolean Expression Q = A+B	Read as A OR B gives NOT Q		

3-input NOR Gate

Symbol	Truth Table			
	C	B	A	Q
	0	0	0	1
	0	0	1	0
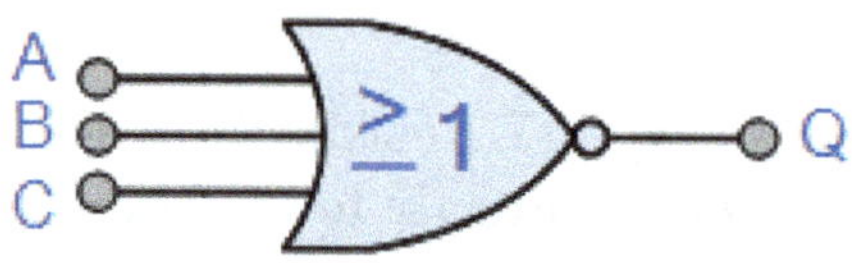	0	1	0	0
	0	1	1	0
	1	0	0	0
3-input NOR Gate	1	0	1	0
	1	1	0	0
	1	1	1	0
Boolean Expression Q = A+B+C	Read as A OR B OR C gives NOT Q			

As with the OR function, the NOR function can also have any number of individual inputs and commercial available NOR Gate IC's are available in standard 2, 3, or 4 input types. If additional inputs are required, then the standard NOR gates can be cascaded together to provide more inputs for example.

A 4-input NOR Function

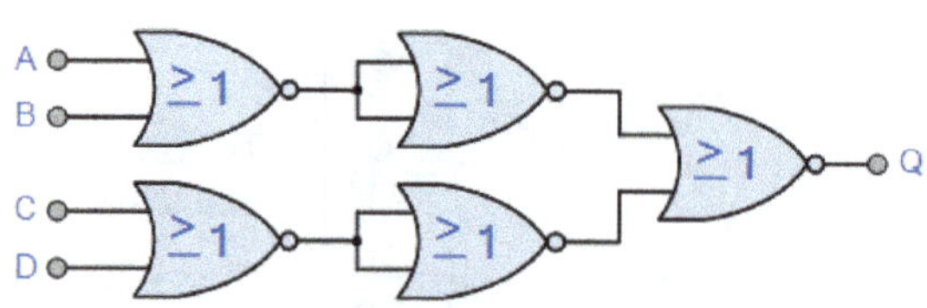

The Boolean Expression for this 4-input NOR gate will therefore be: $Q = \overline{A+B+C+D}$

If the number of inputs required is an odd number of inputs any "unused" inputs can be held LOW by connecting them directly to ground using suitable "Pull-down" resistors.

The Logic NOR Gate function is sometimes known as the Pierce Function and is denoted by a downwards arrow operator as shown, A↓B.

The "Universal" NOR Gate

Like the NAND gate seen in the last topic, the NOR gate can also be classed as a "Universal" type gate. NOR gates can be used to produce any other type of logic gate function just like the NAND gate and by connecting them together in various combinations the three basic gate types of AND, OR and NOT function can be formed using only NOR gates, for example.

Various Logic Gates using only NOR Gates

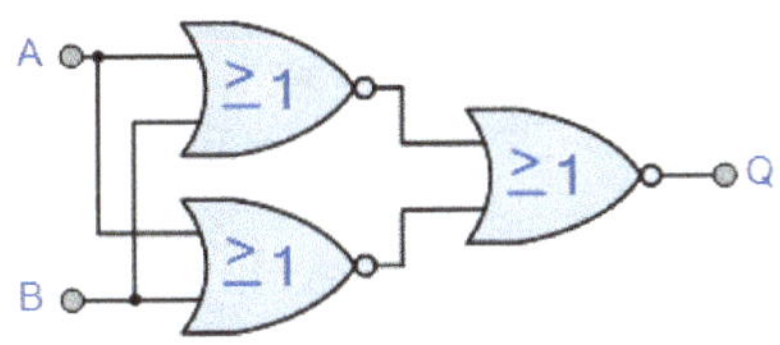

2-input OR Gate

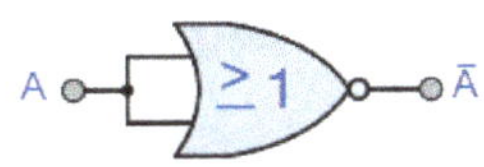

NOT Gate

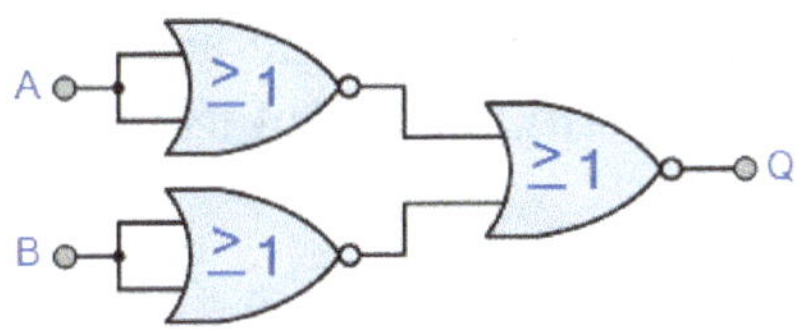

2-input AND Gate

As well as the three common types above, Exclusive-OR, Exclusive-NOR and standard NOR gates can also be formed using just individual NOR gates.

Commonly available digital logic NOR gate IC's include:

TTL Logic NOR Gates.

- 74LS02 Quad 2-input
- 74LS27 Triple 3-input
- 74LS260 Dual 4-input

CMOS Logic NOR Gates.

- CD4001 Quad 2-input
- CD4025 Triple 3-input
- CD4002 Dual 4-input

XOR Gate

The Exclusive-OR logic function is a very useful circuit that can be used in many different types of computational circuits.

The Digital Logic "Exclusive-OR" Gate

2-input Ex-OR Gate

<table>
<tr><th>Symbol</th><th colspan="3">Truth Table</th></tr>
<tr><td rowspan="5">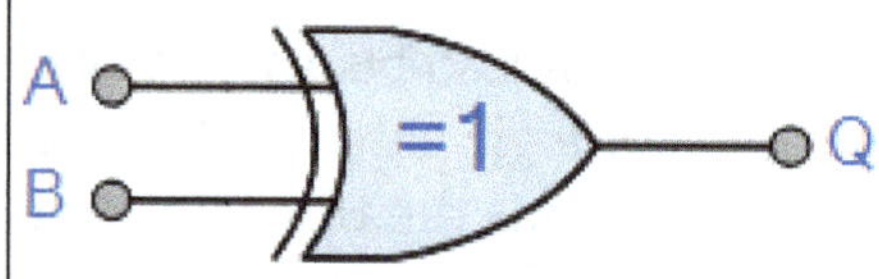

2-input Ex-OR Gate</td><td>B</td><td>A</td><td>Q</td></tr>
<tr><td>0</td><td>0</td><td>0</td></tr>
<tr><td>0</td><td>1</td><td>1</td></tr>
<tr><td>1</td><td>0</td><td>1</td></tr>
<tr><td>1</td><td>1</td><td>0</td></tr>
<tr><td>Boolean Expression Q = A ⊕ B</td><td colspan="3">A OR B but NOT BOTH gives Q</td></tr>
</table>

Giving the Boolean expression of: Q = AB + AB

The truth table above shows that the output of an Exclusive-OR gate ONLY goes "HIGH" when both of its two input terminals are at "DIFFERENT" logic levels with respect to each other. If these two inputs, A and B are both at logic level "1" or both at logic level "0" the output is a "0" making the gate an "odd but not the even gate". In other words, the output is "1" when there are an odd number of 1's in the inputs.

This ability of the *Exclusive-OR gate* to compare two logic levels and produce an output value dependent upon the input condition is very useful in computational logic circuits as it gives us the following Boolean expression of:

$$Q = (A \oplus B) = A.B + A.B$$

The logic function implemented by a 2-input Ex-OR is given as either: "A OR B but NOT both" will give an output at Q. In general, an Ex-OR gate will give an output value of logic "1" ONLY when there are an ODD number of 1's on the inputs to the gate, if the two numbers are equal, the output is "0".

Then an Ex-OR function with more than two inputs is called an "odd function" or modulo-2-sum (Mod-2-SUM), not an Ex-OR. This description can be expanded to apply to any number of individual inputs as shown below for a 3-input Ex-OR gate.

3-input Ex-OR Gate

<table>
<tr><th>Symbol</th><th colspan="4">Truth Table</th></tr>
<tr><td rowspan="9">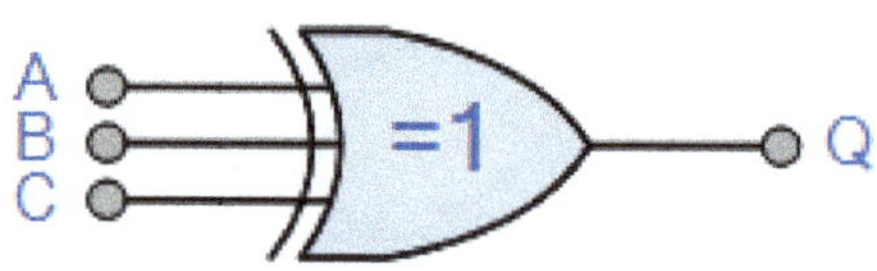

3-input Ex-OR Gate</td><td>C</td><td>B</td><td>A</td><td>Q</td></tr>
<tr><td>0</td><td>0</td><td>0</td><td>0</td></tr>
<tr><td>0</td><td>0</td><td>1</td><td>1</td></tr>
<tr><td>0</td><td>1</td><td>0</td><td>1</td></tr>
<tr><td>0</td><td>1</td><td>1</td><td>0</td></tr>
<tr><td>1</td><td>0</td><td>0</td><td>1</td></tr>
<tr><td>1</td><td>0</td><td>1</td><td>0</td></tr>
<tr><td>1</td><td>1</td><td>0</td><td>0</td></tr>
<tr><td>1</td><td>1</td><td>1</td><td>1</td></tr>
<tr><td>Boolean Expression Q = A ⊕ B ⊕ C</td><td colspan="4">"Any ODD Number of Inputs" gives Q</td></tr>
</table>

Giving the Boolean expression of: Q = ABC + ABC + ABC + ABC

The symbol used to denote an Exclusive-OR odd function is slightly different to that for the standard Inclusive-OR Gate. The logic or Boolean expression given for a logic OR gate is that of logical addition which is denoted by a standard plus sign.

The symbol used to describe the Boolean expression for an Exclusive-OR function is a plus sign, (+) within a circle (O). This exclusive-OR symbol also represents the mathematical "direct sum of sub-objects" expression, with the resulting symbol for an *Exclusive-OR* function being given as: (⊕).

the Ex-OR function is not a basic logic gate but a combination of different logic gates connected together. Using the 2-input truth table above, we can expand the Ex-OR function to: $(A+B).(\overline{A.B})$ which means that we can realise this new expression using the following individual gates.

Ex-OR Gate Equivalent Circuit

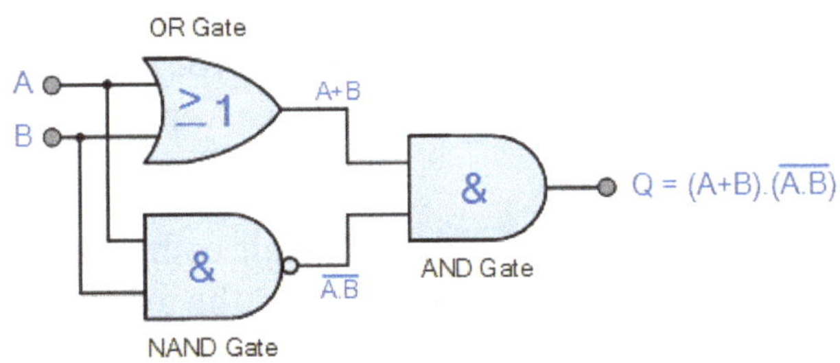

One of the main disadvantages of implementing the Ex-OR function above is that it contains three different types logic gates OR, NAND and finally AND within its design. One easier way of producing the Ex-OR function from a single gate is to use our old favourite the NAND gate as shown above.

Ex-OR Function Realisation using NAND gates

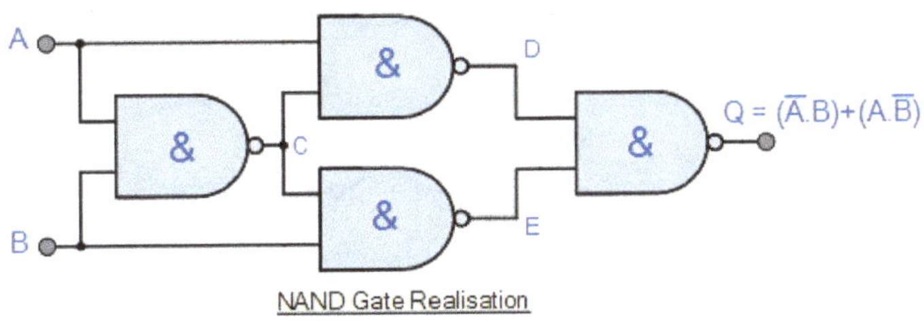

Exclusive-OR Gates are used mainly to build circuits that perform arithmetic operations and calculations especially Adders and Half-Adders as they can provide a "carry-bit" function or as a controlled inverter, where one input passes the binary data and the other input is supplied with a control signal.

Commonly available digital logic Exclusive-OR gate IC's include:

TTL Logic Ex-OR Gates.

- 74LS86 Quad 2-input

CMOS Logic Ex-OR Gates.

- CD4030 Quad 2-input

7486 Quad 2-input Exclusive-OR Gate

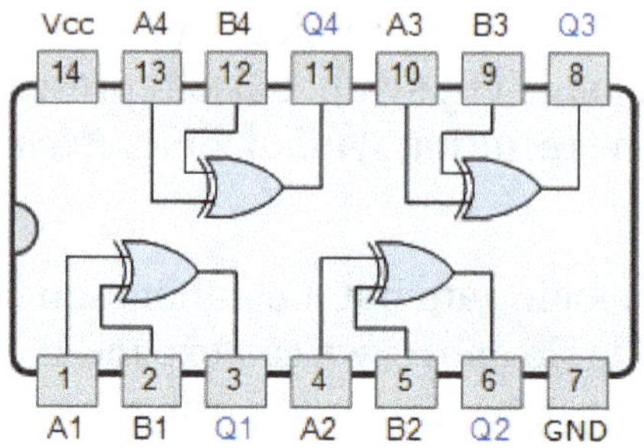

The Exclusive-OR logic function is a very useful circuit that can be used in many different types of computational circuits. Although not a basic logic gate in its own right, its usefulness and versatility has turned it into a standard logical function complete with its own Boolean expression, operator and symbol. The *Exclusive-OR Gate* is widely available as a standard quad two-input 74LS86 TTL gate or the 4030B CMOS package.

One of its most commonly used applications is as a basic logic comparator which produces a logic "1" output when its two input bits are not equal. Because of this, the exclusive-OR gate has an inequality status being known as an odd function. In order to compare numbers that contain two or more bits, additional exclusive-OR gates are needed with the 74LS85 logic comparator being 4-bits wide.

Exclusive-NOR Gate

The operation of Exclusive NOR gate is reciprocal to the Exclusive OR gate's operation. As the basic AND gate implements the Boolean addition, while OR gate implements Boolean multiplication and NOT implements inversion function, there is no such Boolean function defined by the XNOR.This can be derived by basic gates (AND, OR,NOT).Below shows the diagram of EX-NOR constructed from EX-OR gate.

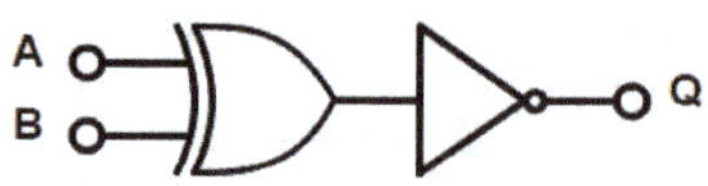

The logic symbol and Boolean expression for the XNOR gate is shown above.

Ex–NOR gate Logic Symbol and Boolean expression

The XNOR logic symbol in IEEE and IEC standards is shown below.

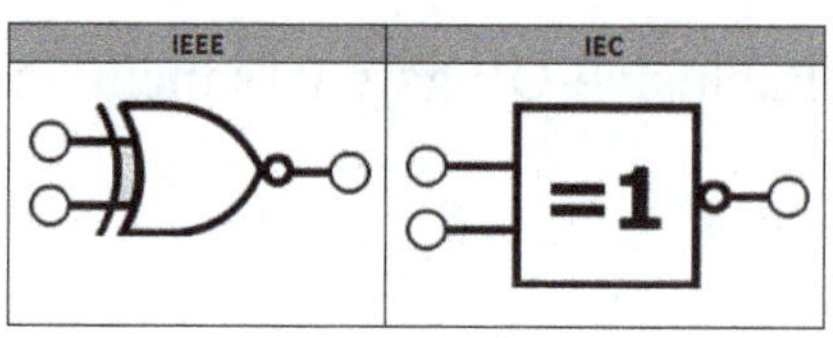

It is a Hybrid gate, the Boolean expression of output of XNOR gate is given as below:

$$A.B+\bar{A}.\bar{B}$$

The XNOR output is represented as:

$$\overline{A\oplus B}$$

XNOR Truth Table

the XNOR gate output will be HIGH when both the inputs are not same and will become LOW for different combination of input.

Input		Outputs
X	Y	Z
0	0	1
0	1	0
1	0	0
1	1	1

X-NOR Gate using Basic Logic Gates

If a specific gate is not available directly, we can design the gate by connecting multiple gates together. An EX-NOR gate can be designed by using basic logic gates like NAND gate and NOR gate.

- Using NOR Gates

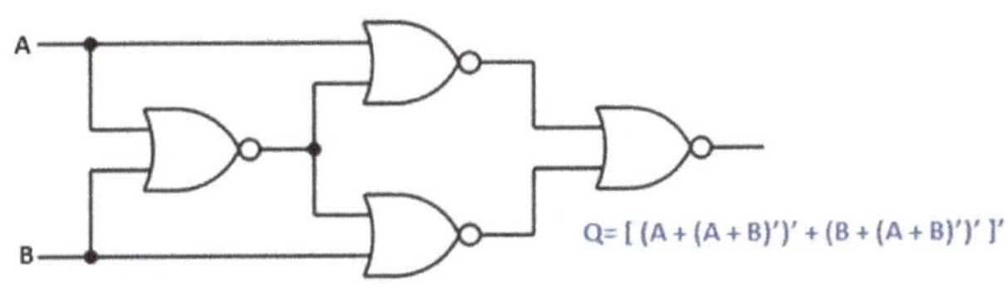

- Using NAND Gates

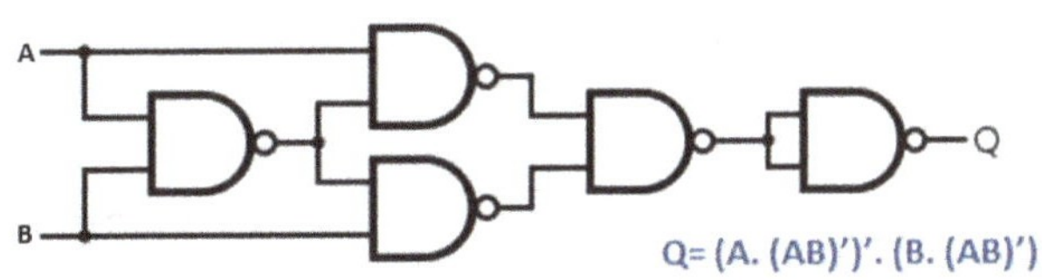

- Using NAND and OR Gates

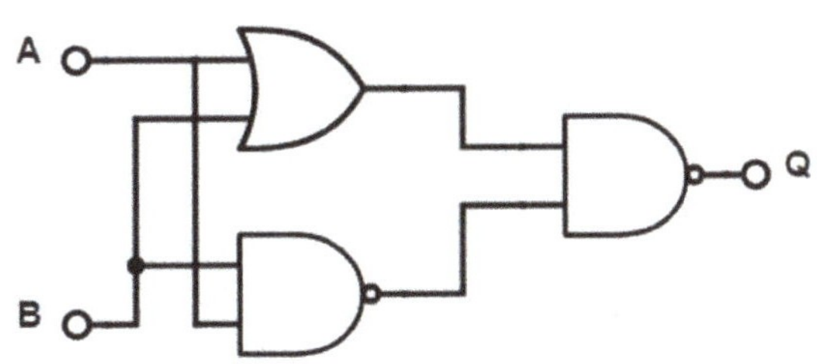

Ex–NOR Gate Pulsed Operation

The pulsed operation of 2 input XOR gate is shown below.

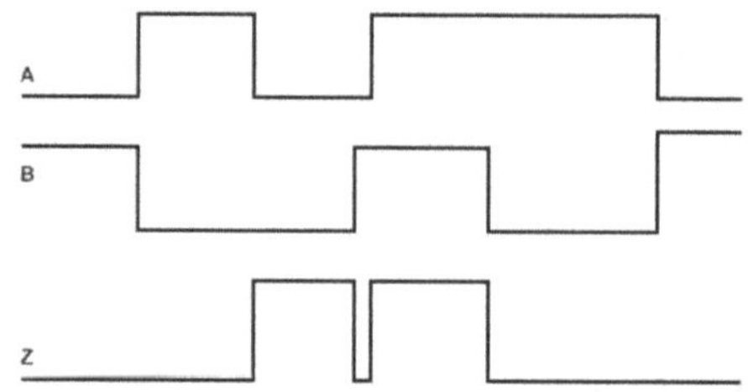

Initially when one of the two inputs of the XNOR gate is low, so the output is at low level. When both the input changes to low level, the XNOR output will rise to HIGH level and again when one of the inputs comes to low level, the output also comes to low level.

3-input Ex-NOR Gate

We can have XOR gate with more than 2 inputs, in some cases. The Boolean function is same for the 3- input XNOR gate is:

$$Q = \overline{A \oplus B \oplus C} = \left(ABC + A'B'C + A'BC' + AB'C'\right)'$$

3-Input Ex-NOR Gate Logic Symbol

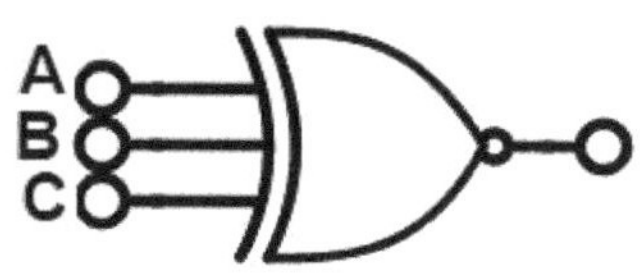

Truth Table

For Exclusive NOR gates, we can have the HIGH input when even numbers of inputs are at HIGH level. So the 3-input XNOR gate is called as "Even functioned OR gate".The truth table and logic symbol for 3-input XNOR gate is given below.

Input			Outputs
W	X	Y	Q=A⊕B⊕C
0	0	0	1
0	0	1	0
0	1	0	0
0	1	1	1
1	0	0	0
1	0	1	1
1	1	0	1
1	1	1	0

Ex-NOR Function Realization using NOR gates

If A and B are inputs and Q is the output of XNOR gate then,

$$Q = A \text{ XNOR } B = (A \text{ XOR } B)'$$

$$Q' = (A \text{ XOR } B) = A'B + AB' = A'B + AB' + AA' + BB' = (A + B)(A' + B')$$

$$Q' = A'(A + B) + B'(A + B)$$

Take compliment,

$$Q = (A'(A + B) + B'(A + B))' = (A'(A + B))'.(B'(A + B))' = (A + (A + B)').(B + (A + B)')$$

Take compliment again,

$$Q' = ((A + (A + B)').(B + (A + B)'))' = (A + (A + B)')' + (B + (A + B)')'$$

Take compliment again,

$$Q = [(A + (A + B)')' + (B + (A + B)')']'$$

Now we can implement the XNOR circuit using NOR gates,

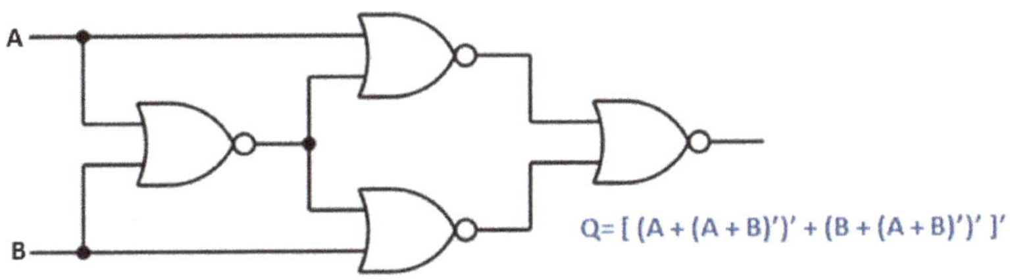

Ex-NOR Function Realization using NAND gates:

$$Q = (A \text{ XOR } B)' = (A'B + AB' = A'B + AB' + AA' + BB')' = ((A + B)(A' + B'))'$$

Now we need to implement this circuit using NAND gates:

$$Q = ((A + B)(AB)')' = (A.(AB)' + B.(AB)')'$$

Taking compliment,

$$Q' = A.(AB)' + B.(AB)' = (A.(AB)').(B.(AB)')$$

Taking compliment again,

$$Q = (A.(AB)')'.(B.(AB)')$$

Now we can implement XNOR gate using NAND gates,

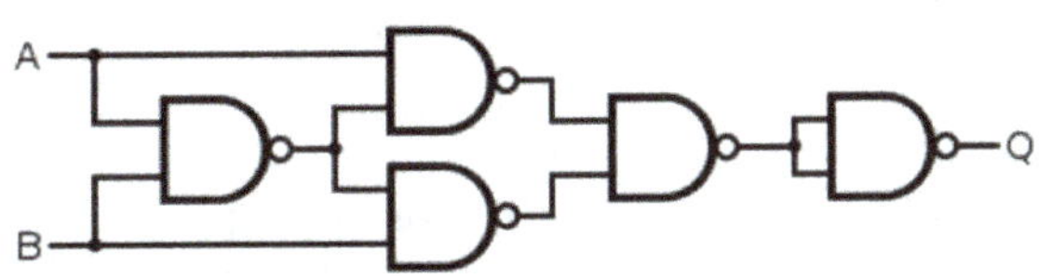

Commonly available TTL and CMOS logic Ex-NOR gate IC's

Some of the most used XNOR logic ICs are listed below:

IC number	purpose
4077	Quad 2-input XNOR gate
74266	quad 2-input XNOR gate with open collector outputs
747266	quad 2-input XNOR gate

74266 Quad 2-input Ex-NOR Gate IC

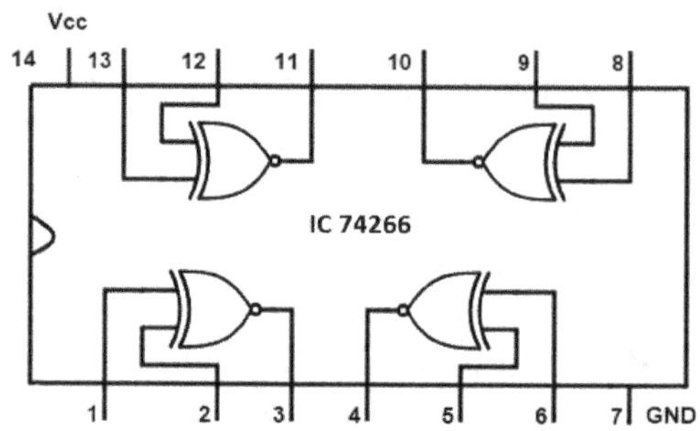

Pin description

Pin Number	Description
1	Input Gate 1
2	Input Gate 1
3	Output Gate 1
4	Output Gate 2
5	Input Gate 2
6	input Gate 2
7	Ground
8	input Gate 3
9	input Gate 3
10	Output Gate 3
11	Output Gate 4
12	Input Gate 4
13	Input Gate 4
14	Positive Supply

Ex-NOR Gate Applications

1. The XNOR logic gates are used in error detecting circuits which are to detect Odd parity or even parity bits in digital data transmission circuits.

2. XNOR gate is mainly used in arithmetic and encryption circuits. This process is the combinational operation of the XOR and XNOR gates, by using 6 transistors for low power application.

This is also used as Heat exchanger tank, which will ring the alarm when the water temperature level goes up or down, to the pre-set level.

Construction

This circuit is used to warm the water. It has two outlets and two float indicators. The float indicators are connected as the inputs of XNOR gate. The output of the XNOR gate is connected to the alarm circuit, which used to alert when the water level changes in the glass beaker. There is a heating element at the bottom of the beaker to heat the certain amount of water.

When the water reaches high level, the excess water will overflow from the upper outlet. When the water comes to low level, then the heating element situated at the bottom is not covered by water so it will burn out. it alerts the when the water level increases or decreases to certain level. To do this, the float switches are to be placed at high level and low level positions of water, inside the tank.

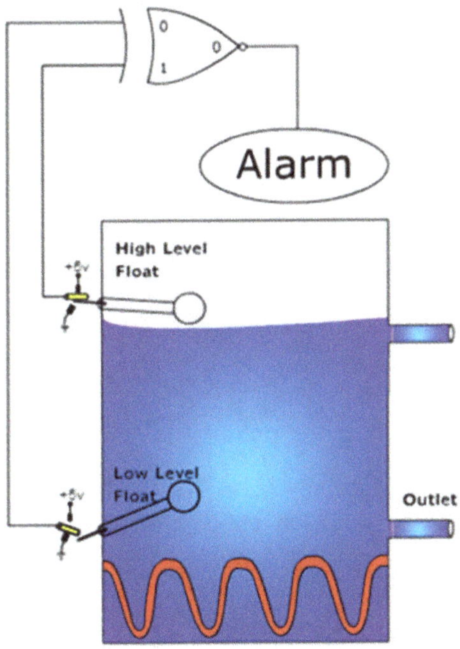

When a float indicator gets in contact with water level, it is pushed upwards. This will cause the float's leaver to disconnect from the armature. At this situation, the XNOR gate's input is 0. If the float indicator is not in contact with the water level, then the leaver is connected to +5 V supply, making the armature active. At this situation, the XNOR gate's input will be HIGH, which is equal to 1. When the 2 leavers are connected to the +5 volts supply or to ground voltage, then the 2 inputs of the XNOR gate is 1, and thus the output of the XNOR gate is high, so that it activates the alarm and alerts the person.

When the water level is between upper floater and lower floater indicator, then logic 1 is applied from upper indicator and logic 0 from lower indicator, to the input of XNOR gate. At this position no alarm will ring as the water level will is at required level. This is how the logic XNOR gate used as a Heat exchanger circuit.

References

- Introduction-analog-electronics: engineersgarage.com, Retrieved 5 May, 2019
- Operational-amplifier-applications: electronicshub.org, Retrieved 7 March, 2019
- The-instrumentation-amplifier, semiconductors: allaboutcircuits.com, Retrieved 15 June, 2019
- Digital_electronics: fandom.com, Retrieved 28 February, 2019
- Logic: electronics-tutorials.ws, Retrieved 23 August, 2019
- Logical-and-gate: electrical4u.com, Retrieved 21 January, 2019
- Not-gate: circuitglobe.com, Retrieved 11 April, 2019
- Nand-gate: electrical4u.com, Retrieved 9 July, 2019
- Exclusive-nor-gate: electronicshub.org, Retrieved 12 February, 2019

Permissions

All chapters in this book are published with permission under the Creative Commons Attribution Share Alike License or equivalent. Every chapter published in this book has been scrutinized by our experts. Their significance has been extensively debated. The topics covered herein carry significant information for a comprehensive understanding. They may even be implemented as practical applications or may be referred to as a beginning point for further studies.

We would like to thank the editorial team for lending their expertise to make the book truly unique. They have played a crucial role in the development of this book. Without their invaluable contributions this book wouldn't have been possible. They have made vital efforts to compile up to date information on the varied aspects of this subject to make this book a valuable addition to the collection of many professionals and students.

This book was conceptualized with the vision of imparting up-to-date and integrated information in this field. To ensure the same, a matchless editorial board was set up. Every individual on the board went through rigorous rounds of assessment to prove their worth. After which they invested a large part of their time researching and compiling the most relevant data for our readers.

The editorial board has been involved in producing this book since its inception. They have spent rigorous hours researching and exploring the diverse topics which have resulted in the successful publishing of this book. They have passed on their knowledge of decades through this book. To expedite this challenging task, the publisher supported the team at every step. A small team of assistant editors was also appointed to further simplify the editing procedure and attain best results for the readers.

Apart from the editorial board, the designing team has also invested a significant amount of their time in understanding the subject and creating the most relevant covers. They scrutinized every image to scout for the most suitable representation of the subject and create an appropriate cover for the book.

The publishing team has been an ardent support to the editorial, designing and production team. Their endless efforts to recruit the best for this project, has resulted in the accomplishment of this book. They are a veteran in the field of academics and their pool of knowledge is as vast as their experience in printing. Their expertise and guidance has proved useful at every step. Their uncompromising quality standards have made this book an exceptional effort. Their encouragement from time to time has been an inspiration for everyone.

The publisher and the editorial board hope that this book will prove to be a valuable piece of knowledge for students, practitioners and scholars across the globe.

Index